SpringerBriefs in Statistics

JSS Research Series in Statistics

The current research of statistics in Japan has expanded in several directions in line with recent trends in academic activities in the area of statistics and statistical sciences over the globe. The core of these research activities in statistics in Japan has been the Japan Statistical Society (JSS). This society, the oldest and largest academic organization for statistics in Japan, was founded in 1931 by a handful of pioneer statisticians and economists and now has a history of about 80 years. Many distinguished scholars have been members, including the influential statistician Hirotugu Akaike, who was a past president of JSS, and the notable mathematician Kiyosi Itô, who was an earlier member of the Institute of Statistical Mathematics (ISM), which has been a closely related organization since the establishment of ISM. The society has two academic journals: the Journal of the Japan Statistical Society (English Series) and the Journal of the Japan Statistical Society (Japanese Series). The membership of JSS consists of researchers, teachers, and professional statisticians in many different fields including mathematics, statistics, engineering, medical sciences, government statistics, economics, business, psychology, education, and many other natural, biological, and social sciences.

The JSS Series of Statistics aims to publish recent results of current research activities in the areas of statistics and statistical sciences in Japan that otherwise would not be available in English; they are complementary to the two JSS academic journals, both English and Japanese. Because the scope of a research paper in academic journals inevitably has become narrowly focused and condensed in recent years, this series is intended to fill the gap between academic research activities and the form of a single academic paper.

The series will be of great interest to a wide audience of researchers, teachers, professional statisticians, and graduate students in many countries who are interested in statistics and statistical sciences, in statistical theory, and in various areas of statistical applications.

More information about this series at http://www.springer.com/series/13497

Yuzo Hosoya · Kosuke Oya
Taro Takimoto · Ryo Kinoshita

Characterizing Interdependencies of Multiple Time Series

Theory and Applications

 Springer

Yuzo Hosoya
Tohoku University
Sendai, Miyagi
Japan

Taro Takimoto
Kyushu University
Fukuoka
Japan

Kosuke Oya
Osaka University
Toyonaka, Osaka
Japan

Ryo Kinoshita
Tokyo Keizai University
Kokubunji, Tokyo
Japan

ISSN 2191-544X   ISSN 2191-5458   (electronic)
SpringerBriefs in Statistics
ISSN 2364-0057   ISSN 2364-0065   (electronic)
JSS Research Series in Statistics
ISBN 978-981-10-6435-7   ISBN 978-981-10-6436-4   (eBook)
DOI 10.1007/978-981-10-6436-4

Library of Congress Control Number: 2017951166

Printed on acid-free paper

This Springer imprint is published by Springer Nature
The registered company is Springer Nature Singapore Pte Ltd.
The registered company address is: 152 Beach Road, #21-01/04 Gateway East, Singapore 189721, Singapore

Preface

This book aims to provide a wide-ranging audience of time-series analysis and econometrics with an introduction to an approach to analyzing vector time-series interdependence. In particular, this book is intended as a practical introduction for real-life applications, emphasizing practical analysis of time series using the proposed causal and allied measures, in particular statistical causal analysis of time-series data produced from real life but not necessarily well-controlled experimental intervention. The book does not go in detail on the (asymptotic) theoretical aspects of statistical time-series analysis; rather, it assumes the standard results. The book's emphasis is on a numerical determination of estimated and large-sample theoretical implications for a practical non-large sample.

Chapter 1 provides an introduction and a brief overview of the literature on empirical causal analysis, and positions this book's approach in this body of literature. The chapter compares a variety of conflicting views or interpretations on how statistical associations can be called causal. In particular, controlled random experiments are compared with observational studies in econometrics. The chapter discusses topics such as the relation of causality and exogeneity in the framework of a simultaneous equation; ancillarity and conditioning in relation to exogeneity; and the relation of causality and prediction improvement in empirical analyses.

Chapter 2 first discusses the formal relation of the Granger and Sims concept of causality and then expounds on the measures of one-way effect, reciprocity, and association. The key approach of this book is the elicitation of a one-way effect component of a supposedly causing series. The measures of interdependence are defined overall as well as frequency-wise quantities in the frequency domain, providing three ways of derivation of the frequency-wise measure. One is based on direct canonical factorization of the spectral density matrix, and the other two are based on distributed-lag representation and innovation orthogonalization. Section 2.5 introduces the overall as well as the frequency-wise measures of reciprocity and association.

To address the problem of third-series involvement, Chap. 3 introduces a partial version of the measures of interdependence. The third-effect elimination we suggest is elimination of the one-way effect component of the third series from a pair of subject-matter series to preserve the inherent feedback structure of the pair. This chapter presents explicit representations of those partial measures and shows how they are numerically evaluated by means of the canonical factorization algorithm by Hosoya and Takimoto (2010). Chapter 3 also shows how the theoretical framework for stationary processes is extended to cointegrated processes.

Using the stationary vector ARMA process, Chap. 4 discusses the statistical estimation of the partial measures of interdependence and testing allied hypotheses. The point is the use of a simulation-based estimation of the covariance matrix of the measure-related statistics and its application to Wald statistics. In Sect. 4.2, we investigate the performance of a small sample of the partial one-way effect measure estimates using Monte Carlo data. To illustrate an analysis of interdependence in the frequency domain, Sect. 4.3 provides an empirical analysis of US interest rates and economic growth data.

Chapter 5 considers the association between structural change in parameters and changes in the causal measure. The chapter proposes an inference method on the association and examines properties of the test statistic by Monte Carlo simulation. Section 5.2 provides a test for measure changes using a sub-sample variance estimation for high-frequency data. Section 5.3 investigates how the proposed test approach works for small sample examples. The two illustrative empirical applications are provided in Sect. 5.4.

Appendix provides technical complements on the concepts of Hilbert space, root modification method, and Whittle likelihood function used in this book.

Regarding the contributors of the chapters, Hosoya wrote Chaps. 1–3; Takimoto and Hosoya wrote Chap. 4, and the computational results were produced by Takimoto; Oya and Kinoshita wrote Chap. 5, and the Appendix was written by Hosoya and Takimoto. Those chapters and the Appendix in the manuscript stage were proofread mainly by Oya. This book is the result of an effective collaboration among the four of us. We have all attended common conferences and meetings, at which some of us made reports on topics related to the contents of this book. We have also held several of our own meetings to discuss the subject matter, computational methods, and the coherence of the content and style of this book.

Sendai, Japan   Yuzo Hosoya
Toyonaka, Japan   Kosuke Oya
Fukuoka, Japan   Taro Takimoto
Kokubunji, Japan   Ryo Kinoshita
July 2017

Acknowledgements

Earlier versions of Chaps. 3–5 were partly presented at the Autumn meeting of the Japanese Economic Society, Tokyo, 2015, at the Japanese Joint Statistical Meeting, Okayama, 2015, at the Recent Progress in Time Series Analysis conferences, Sendai, 2015 and 2016, and so on. We are grateful for helpful comments by participants on those occasions. Parts of the experimental results in this research were obtained using supercomputing resources at Cyberscience Center, Tohoku University. The research is partially supported by JSPS Grant-in-Aid for Scientific Research (C) 22530211 for Yuzo Hosoya, (B) 16H03605 for Kosuke Oya, and (C) 26380271 for Taro Takimoto and the Grant-in-Aid for Young Scientists (B) 16K17103 for Ryo Kinoshita. Finally, we would like to thank Professor Naoto Kunitomo who gave us an opportunity to write this book.

Contents

Chapter 1
Introduction

Abstract In advance of focusing in subsequent chapters on the main theme of the measures of interdependency, Chap. 1 provides a brief overview of the literature on empirical causal analysis and places the theme in a broader perspective, comparing a variety of conflicting views on how certain statistical associations can be viewed as causal. Among others, alluded to is the field experiment model of detecting causal effects by Neyman (1923) and its reliance on a counterfactual assumption. Controlled random experiments are compared with observational studies in econometrics. The concepts of causality and exogeneity in the framework of the simultaneous equation are discussed. Specifically, ancillarity and conditioning in statistical inferences are explained and their relation to exogeneity is expounded. A preliminary concept of Granger causality is introduced, and the role of prediction improvement in empirical analyses is emphasized.

Keywords Ancillarity · Concept of causality · Conditional inference · Controlled random experiment · Cowles approach · Endogenous variable · Exogenous variable · Granger causality · Marshall's causal effect · Neyman model of field experiment · Structural equation model

1.1 On Empirical Causality

The proposition that the empirically observed association of variables does not necessarily imply a causal relation among them is taken as commonplace. The decision that the occurrence of an event is the cause of the occurrence of another event has more importance than merely reporting the observed association between them because the former sometimes accompanies serious consequences involving liability. The empirical identification of causal relations thus seems to frequently demand extra consideration beyond the determination of the joint distribution or the conditional distribution of the variables involved. Hill (1965) mainly focused on epidemiological studies and listed conditions of statistical "causal" associations that are (1) conspicuously strong associations, (2) repeated observations of consistent results, (3) specificities of associations, (4) temporalities of associations, (5) clear dose–response relationships, (6)

© The Author(s) 2017

Y. Hosoya et al., *Characterizing Interdependencies of Multiple Time Series*,
JSS Research Series in Statistics, DOI 10.1007/978-981-10-6436-4_1

plausibilities in view of the state of knowledge of the day, (7) in coherence with existing evidence, (8) experimental evidence as the strongest support, and (9) analogies with established causal inferences. For broader applicability, Cox and Wermuth (2001, p. 70) elaborated on Hill's conditions for statistical causal dependency as follows: (1) availability of an a priori subject-matter explanation, (2) retrospectively convincing subject-matter explanation, (3) conspicuously large effect, (4) natural monotone dependency of effect with levels of the explanatory variable in question, (5) independent studies producing the repeated effect, (6) no major interactions with attributes of the response variable in question, and (7) dependence as a consequence of a massive intervention in the system.

Although some statistical interdependence is symmetric, such as correlation, intrinsically directional dependence in nature also exists, as discussed in Chap. 2. One event being antecedent to another associated event is an important criterion for establishing causality between associated events. If a treatment follows an effect, it cannot be included as a cause. This book intends to propound the concept of a one-way effect to address temporally directed dependence. In contrast, to explore causality in cross-sectional data, some other characterizations, as Hill (1965) proposed, need to be implemented.

Another common understanding among empirical researchers is that the results obtained from controlled randomized experiments generally surpass comparable findings from non-experimental observations for detecting causality. An effect caused by a treatment can be isolated by introducing randomization and control; moreover, well-designed controlled randomized experiments enable the quantitative evaluation of the sampling distribution of the effect estimate. In contrast, observational studies devoid of controlled interventions do not in general realize such isolation of the subject-matter causal relation. Historical empirical economic analyses lacking controlled intervention have represented efforts to manage this inherent shortcoming using the sophistication of fitted models and statistical analyses.

Experimentalists usually limit the use of causal effects only in observed consequences of experimenters' active interventions. They believe that non-experimental observational results should be evaluated in view of the departure from or similarity with comparable controlled experiments, and they use such simulations or quasi-experimental terms to differentiate them from proper controlled experiments. From the standpoint of experimentalists, such noninterventional variables as intrinsic attributes of observation units are not included in the cause variable. A typical example of a non-cause variable is gender. Differences in sex cannot be a cause of wage differentials because observing the wages of units under a change in sex is impossible. [See, for example, Holland (1986).]

Neyman (1923) illustrated the detection of a causal effect in an agricultural field experiment. Suppose that a field is divided into m plots of equal area and that $U_1, U_2, \ldots, U_m$ are true yields of a particular crop variety. Neyman (1923) used the probability model of random allocation of v varieties by random sampling from urns without replacement. Suppose that we have v urns and the i-th urn contains m balls corresponding to m plots. On each ball is written one of $U_{i1}, U_{i2}, \ldots, U_{ik}, \ldots, U_{im}$. The ball of U_{ik} being drawn from the i-th urn indicates that the yield of the i-th variety allocated on the k-th plot is U_{ik}. The quantity $\bar{U}_i = \sum_{k=1}^{m} U_{ik}/m$ is the true yield of the i-th variety from the entire field. The balls are drawn from the urns without replacement. Moreover, if a ball is drawn randomly from an urn, all balls with the same plot number are withdrawn from all other urns. In other words, no more than one variety is allocated on one plot. The final objective is a comparison of $\bar{U}_1, \bar{U}_2, \ldots, \bar{U}_v$ or a comparison of those true values using a sample set. The causal effect attributable to variety is represented by the differences $\bar{U}_i - \bar{U}_j$ $(i \neq j)$; however, because different varieties i and j are not applied to the same plot, they are not observed. Let X_{ik} be the result of the k-th drawing without replacement from the i-th urn. The quantity U_{ik} is latent and not necessarily observable; however, X_{ik} is the observed value and we have $X_{ik} = U_{ik}$ for the observed X_{ik}. The effects of the varieties are observed as $\bar{X}_i - \bar{X}_j$, and to calculate the variance of this difference, we need knowledge of the correlation coefficient

$$r = \frac{(1/m) \sum_{k=1}^{m} U_{ik} U_{jk} - \bar{U}_i \bar{U}_j}{\sigma_{U_i} \sigma_{U_j}}.$$

However, because different varieties are not compared in the same plot, the value is not directly obtained from the sample. Neyman (1923) proposed expediently $r = 1$ in the paper. Neyman's experimental model is summarized as follows.

1. The yield on the k-th plot of the j-th variety for all k and j is latently given beforehand. The absence of a selection bias and independence between causal effects and the allocation of treatments are modeled.

2. Not more than one variety is allotted in one plot; therefore, to enable estimation of sample variability of the estimate of a causal effect, Neyman relied on the assumption $r = 1$.

3. Yields are written on balls. Such a temporal characterization as a cause preceding an effect is not modeled.

Propositions involving non-observable quantities, such as item 2, are said to be counterfactual. Neyman's model of causality contains characteristically counterfactual elements. In the paper, Neyman used a nonparametric model and does not use the conventional regression model to represent an interventional experiment and the consequences. See Freedman (2010) for critical essays against the use of the conventional regression model and the accompanied statistical inference for non-experimental data.

Although controlled randomized experiments are generally supposed to be superior to observational studies, they are not immune to difficulties. Cox and Wermuth (2001, p. 67) pointed out, regarding such serious difficulties, possible interactive effects with unobserved explanatory variables and unanticipated future interventions that remain unnoticed when data are collected. Zeisel and Kaye (1997) made the following critical remarks on controlled experiments.

A basic limitation of all experiments is that they are conducted at a particular place, at a particular time, and under particular conditions specified by their experimental design. Similar to a searchlight, they provide powerful illumination; however, their lights fall on a tiny area. A well-designed experiment reveals how one variable responds to changes in variables under the experimenter's control, whereas all other relevant variables are held constant (or subject only to random fluctuations) (p. 5). The basic simple structure of the controlled experiment may convey the impression that, properly conducted, it will yield a vast amount of knowledge. However, the yield is limited because of, primarily, the narrow scope of all experiments. ⋯ To extrapolate from these conditions always raises questions (p. 25).

We may summarize their view as follows. Extrapolation relies on the crucial assumption that in-sample conclusions obtained from available data are generalizable to the out-sample; however, the experiment by itself does not justify the assumption. However well designed a controlled experiment is, the possibility of lurking effects is not completely ruled out. Extra knowledge is necessary to justify extrapolation to real-life data. Holland (1986) maintained that the concept of causal relation should be exclusively used for interventional experimental results; however, the passivity of noninterventional observations cannot be completely removed from controlled experiments. Concepts that are too narrowly defined limit their application. Instead of complying with rigid formal rules of intervention, an examination from such broad perspectives, as suggested by Hill (1965) and Cox and Wermuth (2001), is appropriate.

We must also notice that the empirical knowledge resulting from an experimental intervention to validate causality should assist in improving the precision of a prediction. Therefore, we cannot ignore Granger's formalization of causality. The detection of causal directions and their extent in a time-series set is of major interest in time-series analysis, and the literature has centered around the causality concept that Granger (1963) introduced as a statistically testable criterion defined in terms of prediction improvement based on the basic assumption that the cause chronologically precedes the effect and the future does not cause the past. Informally, given a pair of time series, one series is said to cause (not to cause) another in the Granger sense if the addition of the causing series to the predictor set helps (does not help) the prediction of the caused series. The experimental testing model of Neyman (1923) does not consider the passage of time for cause to produce its effect. In contrast, in Granger causality, the temporal antecedent is a crucial factor for a variable to be a candidate of a cause. Section 2.2 formally discusses the definitions of the Granger and Sims causality, and Sect. 2.3 proposes the quantitative measurement of Granger causality.

1.2   Causality in Economic Analysis

In social life, causal relations are often pursued to determine the responsibility of undesirable consequences, as in product liability litigation. In science, to explain a sequence of occurrences of phenomena using causal chains is taken as a descriptive protocol. Marshall (1930) placed causal economic analyses on equal footing with other scientific studies.

1. Economists study like other scientists the effects which will be produced by certain causes subject to the condition that *other things equal*, and that the causes are able to work out their effects undisturbed.

2. In almost every scientific doctrine, we find a proviso that states to the effect that all other factors are equal: the action of the causes in question is supposed to be isolated; certain effects are attributed to the action of causes, but only *on the hypothesis* that no cause is permitted to enter except those distinctly allowed for.

3. That time must be allowed to causes to produce their effects is a source of significant difficulty in economics. Meanwhile, the material on which they work and perhaps even the causes themselves may have changed, and the tendencies being described do not have a sufficiently "long run" in which to work themselves out fully.

4. Although economic analysis and general reasoning have broad applications, every age and every country has its own problems, and every change in social condition is likely to require a new development of economic doctrines.

Regarding isolation of the causing variables in question, Marshall seemed to indicate the hypothetical manipulation of the variables involved by setting other variables unchanged in theoretical setups. Regarding point 3 on the difficulties accompanying the passage of time for the cause to work out when disturbing effects do not intervene, Marshall seemed to indicate an empirical observation of causality. [These citations are from pages 36 to 37 of the 1930 edition of *Principles of Economics*, which is a reprint of the 8th edition (1920).]

Regarding prior empirical economic analysis, Yule (1899) inquired into the causes of change in pauperism during a 10-year period using a cross-sectional regression analysis of 600 regions called unions based on census data for 1881 and 1891. The business cycle study by Tinbergen (1939) is a systematic application of statistical regression analysis for empirical economic analysis. Keynes (1939) immediately responded by criticizing Tinbergen's use of the statistical method on non-experimental economic data whose generating processes are obscure and not well defined. The Keynes' criticism developed into a dispute between them; see Tinbergen (1940) and Keynes (1940). In contrast to Keynes' dismissal of inferring causality from non-experimental observations, he was positive. Tinbergen's positive approach was inherited by the Cowles Commission's empirical analysis and modern econometrics. However, their conflict of views remains as the essential opposition up to now.

Wold (1956, pp. 36–37) presented a view of causality that parallels that of Marshall. Comparing the experimental method with artichoke cookery, Wold emphasized the importance of disentangling a complex causal problem into separate simpler ones; see Wold (1956, p. 36). He suggested that the devices pertinent for that purpose are:

1. Varying one or a few of the controlled causal factors at a time, keeping the others constant;

2. Neutralizing the effect of uncontrolled factors by randomization; and,

3. Arranging the effects of controlled variables to make feasible the application of a linear regression model.

The device 1 above formally corresponds to Marshall's detection of a causal effect under the condition that all other factors are equal. However, a substantial difference exists in that Wold referred to an experimental procedure, whereas Marshall referred to manipulation of variables in a model. In observational studies devoid of the random assignment of treatments, device 2 is difficult to carry out. Lacking device 2, the assumption of independence between explanatory variables and the disturbances is infeasible. Additionally, device 3 is hardly feasible if the explanatory variables are not controllable. Controlled experiments decompose causal relations into simple relations, but observational studies including most economic analyses that lack such devices must typically start from a comprehensive model involving many latent causal relations.

Although economics is supposed to be an empirical science, as are other social sciences, little effort has been made to empirically test economic theories. Rather dominant is the preconceived notion that economic phenomena are theoretically explained by the subjective reasoning and rational behavior of economic agents. Empirical evidence is often shown only in tables and graphs, whereas formal statistical testing of economic theories is not given definitive roles. From the viewpoint of economists, economic models should be deduced from theoretical assumptions. Hence, econometricians' roles should be limited to statistical estimations of parameters of theoretical models. [See Sutton (2000) for a critical essay on economists' attitudes toward empirical economic analyses.] Cox and Wermuth (2001, p. 69) propounded the view that any normal science assumes a certain series of causal chains, from cause variables to response variables accompanying intermediate variables, and that the role of statistical analysis is to test this assumption. However, many authors including Keynes (1939, 40) and Hicks (1979) do not seem to acknowledge testing economic causality using non-experimental observations for empirical confirmation.

1.3 Empirical Economic Models

Researchers assembled in the Cowls Commission founded in 1935 systematized a method for analyzing economic phenomena using observed data. [Later, in 1955, it was renamed the Cowles Foundation and was relocated to Yale University.] Haavelmo (1944), a member of the Commission, proposed economic modeling and statistical inference using a multivariate statistical method. His basic concept is that observed economic data variations are representable by multivariate stochastic models, and economic relations are representable by (multivariate) regression models. Another member, Marschak (1953), suggested the use of structural equation systems to evaluate policy effects. The Commission's contributions are the formalization of identifiability conditions and the development of statistical estimation methods of over-identified equation parameters. They remain until the present as a research paradigm for empirical economic research.

1.3.1   The Cowles Approach

The main purpose of empirical economics is to detect and utilize relationships among economic variables that are invariable under changes in external conditions. Subject-matter economic variables are termed endogenous variables, and the observable external conditions are termed exogenous variables. To model the causal assumption of endogenous variables, the Cowles approach uses a linear simultaneous equation system represented by

$$Au(t) + Bv(t) = \eta(t), \quad t = 1, \ldots, T, \tag{1.1}$$

where A and B are $p \times p$, $p \times q$ matrices, respectively; $u(t)$ is a p-vector endogenous variable; $v(t)$ is a q-vector exogenous variable; $\eta(t)$ is a p-vector disturbance term with mean 0 and covariance matrix Ω; and $u(t)$ and $v(t)$ are observable, whereas $\eta(t)$ is an unobservable random vector. The values $v(t)$ and $\eta(t)$ are supposed to be determined outside the system (1.1), whereas the value of the endogenous variable $u(t)$ is determined by the system given $v(t)$ and $\eta(t)$. Therefore, the endogenous variables are correlated with the disturbance terms, while the exogenous variables are not correlated with (or independent from) the disturbance terms. The simultaneous equation system (1.1) is termed a structural form or structural equation. When the coefficient matrix A is non-singular, (1.1) determines the value of $u(t)$ and the structural form is said to be complete. If the system (1.1) is complete, it is solved for $u(t)$ as

$$u(t) = \Phi v(t) + \varepsilon(t). \tag{1.2}$$

The Eq. (1.2) is termed the reduced form of (1.1), where $\Phi = -A^{-1}B$ and the disturbance term $\varepsilon(t)$ has mean 0 and covariance matrix $\Sigma = A^{-1}\Omega(A^{-1})'$.

Statistical estimation and testing are conduced generally using the likelihood function (or its substitute) derived from the reduced form (1.2) and on a certain set of a priori restrictions imposed on the structural form (1.1). Once the assumption of the *i.i.d.* (or *i.i.d.* normal) white noise of the process $\{\eta(t)\}$ and the a priori identifiability restrictions are imposed, the Cowles approach based on the pair of forms (1.1) and (1.2) is reduced to special cases of statistical multivariate analysis.

Marshall's causality placed in this framework is paraphrased as a characterization of how a set of endogenous variables is affected by variations in a set of endogenous and/or exogenous variables when the other variables are fixed. In particular, the extent of the causal effect from exogenous variables to endogenous variables is determined by the coefficient Φ, and the effect is said to be a multiplier effect. The determination of Marshall's causality relies on the identifiability of the related parameters. In this respect, although the reduced-form coefficient Φ is in most cases identifiable, structural parameters are not uniquely derived from the reduced-form

parameters and a sufficient number of restrictions (mostly zero restrictions on parameters) are required to be imposed on each equation in the system (1.1) for each to
be identifiable. An evident contribution of the Cowles Commission is that it made
explicit the identification problem and the conditions for identifiability. However, the
imposition of those identifiability conditions in modeling real economic data by a
priori reason or out of expediency often conflicts with both empirical and theoretical
plausibility; see Sims (1980).

During the 1950s and 1960s, statistical inference based on simultaneous equation
models was widely used for empirical macroeconomic analyses. In contrast, in the
1960s, empirical economic time-series analysis was developed independent from
the Cowles program; see Granger-Hatanaka (1964). In retrospect, the transition of
empirical economic analysis from comparative statics based on multivariate statistical analysis to dynamic econometric analysis based on stationary stochastic theory
and statistical time-series analysis overlaps with the period of the shift of the empirical macroeconomic research paradigm. To summarize the retrospective remarks in
Klein (1981) regarding the circumstance in those days of transition:

1. Application of highly sophisticated statistical methods, in particular simultaneous esti-
    mation methods, did not improve prediction precision.
2. A distributed-lag approach employing intensive computer inputs contributed to improv-
    ing the understanding of model dynamics.

Remark 1.1   We may say it was a manifold twist in the history of macroeconomics
that the criticism against macroeconometric analyses based on the IS-LM model
complemented by the Phillips curve during the 1970s and 1980s was taken as a
synonym of the criticism against Keynes economics. In the first place, Keynes was
critical of Hicks' IS-LM modeling, maintaining that it did not reflect correctly his
idea. Moreover, the original theory Keynes propounded did not contain the Phillips
curve. Second, both Keynes (1930, 1940) and Hicks (1979) objected to applying
statistical methods to non-experimental data such as macroeconomic data.

1.3.2   Economic Time-Series Models

If predetermined endogenous variables are included in a reduced form (1.2) in the
form of a distributed lag, the reduced form is represented by the multivariate autoregressive (VAR) model involving exogenous variables: namely,

$$u(t) = \sum_{j=1}^{a} A[j]u(t - j) + \Phi v(t) + \varepsilon(t) \tag{1.3}$$

where the disturbance $\{\varepsilon(t)\}$ is a white noise process. In the sequel a, b, c of upper
limits of summations denote lag orders and hence positive integers. To take into

account the lagged effects of the exogenous variable, we have the autoregressive distributed-lag (ARDL) model represented by

$$u(t) = \sum_{j=1}^{a} A[j]u(t-j) + \sum_{k=1}^{c} \Phi[k]v(t-k) + \varepsilon(t). \tag{1.4}$$

To allow weak dependence on the disturbance term, we can introduce the moving average process for it, and then, we have the following multivariate autoregressive-moving average (VARMA) model (containing exogenous variable $v(t)$)

$$u(t) = \sum_{j=1}^{a} A[j]u(t-j) + \Phi v(t) + \varepsilon(t) + \sum_{k=1}^{b} B[k]\varepsilon(t-k). \tag{1.5}$$

Recent studies (e.g., Athanasopoulos and Vahid (2008)) suggested that the ARMA model has greater macroeconomic prediction precision than the AR model. If $\{u(t)\}$ is generated by (1.3), (1.4), or (1.5), and the mean-corrected process $\{u(t) - Eu(t)\}$ is stationary, statistical inference of stationary time-series analysis is directly applied for inference on the parameters of those models.

Giving up the distinction of endo and exogeneity, Heckman (2000) suggested the following determination of $\{u(t), v(t)\}$:

$$\begin{bmatrix} A & B \\ 0 & I \end{bmatrix} \begin{bmatrix} u(t) \\ v(t) \end{bmatrix} \tag{1.6}$$

$$= \sum_{j=1}^{a} \begin{bmatrix} A_{11}[j] & A_{12}[j] \\ A_{21}[j] & A_{22}[j] \end{bmatrix} \begin{bmatrix} u(t-j) \\ v(t-j) \end{bmatrix} + \begin{bmatrix} \varepsilon_1(t) \\ \varepsilon_2(t) \end{bmatrix} + \sum_{k=1}^{b} \begin{bmatrix} B_{11}[k] & B_{12}[k] \\ 0 & B_{22}[k] \end{bmatrix} \begin{bmatrix} \varepsilon_1(t-k) \\ \varepsilon_2(t-k) \end{bmatrix},$$

where $\varepsilon_1(t-k), \varepsilon_2(t-k), k = 0, 1, \ldots, b$ are, respectively, p and q-vector white noise, and we assume $\{\varepsilon_1(t)\}$ is orthogonal to $\{\varepsilon_2(t)\}$. Combining the characteristics of the Cowles and the time-series models, the Eq. (1.6) provides an explicit embodiment of Granger's one-way causality from $\{v(t)\}$ to $\{u(t)\}$.

When the characteristic equation

$$\det\left(I_p - \sum_{j=1}^{a} A[j]z^j\right) = 0$$

of a process generated by (1.3) or (1.5) has a unit root $z = 1$, the process is said to be a unit-root process. A typical economic mechanism produces a stochastic trend. Suppose that a time series $\{W_T(t)\}$ is generated by the VARMA model (1.5), and suppose that there is among the economic quantity $W_T(t)$ a long-term equilibrium relation $\beta' W_T(t) = \mu$ (where β is a $p \times r$ matrix with rank r and μ is a r-vector). The long-term equilibrium implies that the relation is dynamically stable. A representation of stability is stochastic stationarity, or more specifically, we assume that $\beta'W_T(t)$ is second-order stationary. A column vector β, which makes $\beta'W_T(t)$ stationary, is termed a cointegration vector. Let $\xi(t)$ be a MA process generated by

$$\xi(t) = \sum_{j=0}^{b} B[j]\varepsilon(t - j),$$

where $B(0) = I_p$ and $\{\varepsilon(t)\}$ represent an orthogonal sequence of random variables with mean 0 and full-rank covariance matrix Ω. The cointegration model for the process $\{W_T(t)\}$ that explicitly represents the adjustment mechanism that prevents departure from the equilibrium is the error correction model given as follows:

$$\Delta(L)W_T(t) = \Pi W_T(t-1) + \sum_{k=1}^{a-1} A[k]\Delta(L)W_T(t-k) + \Gamma g_T(t) + \xi(t), \ t = 1, \cdots, T,$$

$$(1.7)$$

where $\Pi = \alpha\beta'$ for $p \times r$ matrices α, β, $\Delta(L) = (1 - L)I_p$, and $g_T(t)$ is a deterministic q-vector involving no unknown parameters, and Γ is a $p \times q$ coefficient parameter. The member $\sum_{k=1}^{a-1} A[k]\Delta(L)W_T(t-k)$ in (1.7) is supposed to be absent if $a = 1$. We assume that the $W_T(t)$ for $-a + 1 \le t \le 0$ are uniformly bounded in probability with respect to T to derive large-sample estimation properties.

We can derive a moving average representation of $\Delta(L)W_T(t)$ and consequently of $W_T(t)$ as follows. Defining $A(z)$ by

$$A(z) \equiv \Delta(z) - \Pi z - \sum_{k=1}^{a-1} A[k]z^k\Delta(z),$$

write the Eq. (1.7) as

$$A(L)W_T(t) = \Gamma g_T(t) + \xi(t).$$

Let β and $(\beta, \beta_\perp)$ be $p \times r$ and $p \times p$ matrices such that $\beta \perp \beta_\perp$ and $rank(\beta, \beta_\perp) = p$, where $0 \le r \le p$. Suppose that $\Pi = \alpha\beta'$ for a full-rank $p \times r$ matrix α. Let $J(z)$ be the block diagonal matrix given by

$$J(z) = \left[\begin{array}{c|c} I_r & 0 \\ \hline 0 & (1-z)I_{p-r} \end{array}\right].$$

Then, we have

$$A(z)(\beta, \beta_\perp) = \left[A(z)\beta, \ \beta_\perp - \sum_{k=1}^{a-1} A[k]z^k\beta_\perp \right] J(z) \equiv C(z)J(z). \quad (1.8)$$

It follows from (1.8) that a necessary and sufficient condition for det $C(z)$ to have all zeros outside the unit circle is that det $A(z)$ has $(p - r)$-multiple unity zeros and

r zeros are outside the unit circle. Denote by $C(z)^\sharp$ the adjugate matrix (*i.e.*, the transposed cofactor matrix) of $C(z)$: namely,

$$C(z)^\sharp C(z) = \{\det\ C(z)\} I_p.$$

Let $H(z)$ be the block diagonal matrix given by

$$H(z) = \left[\begin{array}{c|c} (1-z)I_r & 0 \\ \hline 0 & I_{p-r} \end{array}\right];$$

then, it follows from (1.8) that

$$H(z)C(z)^\sharp A(z)(\beta, \beta_\perp) = H(z)C(z)^\sharp C(z)J(z) = (\det\ C(z))\Delta(z).$$

Consequently, because $\Delta(z)$ and $(\beta, \beta_\perp)^{-1}$ are commutable, we have

$$H(z)C(z)^\sharp A(z) = \{\det\ C(z)\}(\beta, \beta_\perp)^{-1}\Delta(z). \tag{1.9}$$

Therefore, if $\det\ C(z)$ has all of the zeros outside the unit circle, we have the relationship

$$\{\det\ C(z)\}^{-1}(\beta, \beta_\perp)H(z)C(z)^\sharp A(z) = \Delta(z). \tag{1.10}$$

The equality (1.10) implies the next theorem, which is said to be the Granger representation theorem; see Johansen (1991).

Theorem 1.1 *If and only if the r zeros of $\det\ A(z)$ are outside the unit circle and the rest are equal to 1, the process $\{W_T(t), t = 1, \cdots, T\}$ has the representation*

$$\Delta(L)W_T(t) = (\beta, \beta_\perp)H(L)C(L)^{-1}\{\Gamma g_T(t) + \xi(t)\}.$$

Remark 1.2 It follows from definition (1.8) that the expansion of $C(z)^{-1}$ has exponentially decaying coefficients.

Remark 1.3 As this proof shows, it is essential for the derivation of representation (1.9) that $\Delta(z)$ and $(\beta, \beta_\perp)^{-1}$ are commutable. Because $J(z)$ and $(\beta, \beta_\perp)^{-1}$ are generally not commutable, the moving average representation of $J(L)W_T(t)$ does not directly follow from the error correction model (1.8) even though the vector $J(L)(\beta, \beta_\perp)^{-1}W_T(t)$ has the moving average representation

$$J(L)(\beta, \beta_\perp)^{-1}W_T(t) = C(L)^{-1}\{\xi(t) + \Gamma g_T(t)\}.$$

1.4   Basic Concepts for Statistical Inference

Evaluation of statistical evidence is ordinarily conducted using a set of observed data in view of probability model(s), where the latter is composed of the triplet of the sample space Ω to which observation x belongs, the parameter space Θ to which the parameter θ of the probability model belongs, and the probability Pr defined on the product space of Ω and Θ. In this section, Pr indicates the probability for the discrete probability model and the probability density for the continuous probability model. The triplet is expressed by $M \equiv (\Omega, \Theta, Pr)$ and M is termed an experiment. An evaluation of the statistical evidence in this framework is about interpreting the evidential content or evidential meaning of the result x in the experiment M. The notation $Ev(M, x)$ is used in the sequel as a short hand writing of the expression "evidential meaning of data x in the presence of experiment M." What it implies is not given in advance, but its role is revealed by stating how the concept is practically used in statistical inference; see Birnbaum (1962) for this framing of statistical inference.

1.4.1   Conditional Inference

Sufficiency of a statistic is the concept that Fisher (1925) introduced in relation to statistical estimation, which he regarded as data reduction. If a certain experimental result is summarized as statistic(s) (e.g., to summarize 50 observations into the sample average), the summary usually accompanies a certain amount of information loss. A statistic being sufficient implies that summarizing into the statistic does not incur loss of information. Given an experiment (Ω, Θ, Pr), a function $t(x)$ of a result x is a sufficient statistic if further knowledge of x does not bring additional information on θ. In terms of a probabilistic expression, $t(x)$ is a sufficient statistic if we have the relation

$$Pr(x|\theta) = Pr(x|t)Pr(t|\theta),$$

for $(x, \theta) \in \Omega \times \Theta$, where $Pr(x|t)$, the conditional probability x of given t, is independent of θ. Denote by $\Omega' = \{t(x) : x \in \Omega\}$ the set of possible values of $t(x)$. The experiment in which $t(x)$ is observed is expressed as (Ω', Θ, Pr'), where Pr' indicates the probability $Pr' = Pr(t|\theta)$. Conditional distribution of any statistic given a sufficient statistic is independent θ. An application of this property is as follows. Suppose that t is a sufficient statistic and we have an estimate $\tilde{\theta}(x)$ of parameter θ. Then let $\tilde{\theta}_1 \equiv E(\tilde{\theta}|t)$ be the conditional expectation of $\tilde{\theta}$ given t. If a loss function $V(\hat{\theta}, \theta)$ is a real-valued convex function with respect to any estimate $\hat{\theta}$ for fixed θ, we have in general the relation

$$E(V(\tilde{\theta}_1, \theta)) \leq E(V(\tilde{\theta}, \theta)). \tag{1.11}$$

Furthermore, the property that the conditional probability of result x given t is pivotal can be used to construct a test when the null hypothesis is a composite.

Inequality (1.11) justifies our construction of an estimate on the basis of the experiment $M' = (\Omega', \Theta, Pr')$ from the decision-theoretic standpoint, whereas the axiom of sufficiency asserts that

$$Ev(M, x) = Ev(M', t),$$

which asserts that there is no difference in evidential content between (M, x) and $(M', t(x))$.

Another concept that Fisher introduced, along with sufficiency, is ancillarity; see Fisher (1934). The next example is attributable to Cox (1958).

Example 1.1 Suppose that we have two experiments M_1 and M_2 that produce the result y, which is normally distributed with mean θ and variances 1 and 4 under M_1 and M_2, respectively. Moreover, we have experiment M_3 whose result z takes 0 or 1, respectively, with probability $1/2$ independently with M_1 and M_2. Let M^* be the mixed experiment in which M_1 is conducted if $z = 0$ and M_2 is conducted if $z = 1$, such that the result of experiment M^* is the pair (z, y). To infer θ, suppose that we conducted experiment M_3 and the result of z was 0. Consequently, experiment M_1 was carried out, and suppose that $y = 1.5$ was observed. Now, the question is whether $Ev(M_1, 1.5) = Ev(M^*, (0, 1.5))$ is the appropriate way to evaluate the evidence. For inference on θ, the actually conduced experiment is M_1 and not M_2. However, if experiment M^* is to be continued, M_2 is to be conducted and we cannot disregard the outcome. From the viewpoint that the result should be interpreted only on the basis of the actually conducted experiment, Cox argues that $(M^*, (0, 1.5))$ should have a same meaning as $(M_1, 1.5)$, which implies that conditional inference based on the result of z is relevant.

Fisher terms the statistic z in the previous example as an ancillary statistic and employs it as a conditioning variable to delimit the sample space in relation to the subject-matter inference problem. For an experiment $M = (\Omega, \Theta, Pr)$ with a discrete probability model, if the probability

$$Pr\{t(x) = t\} = \sum_{t(x)=t} Pr(x)$$

of $t(x)$ taking t is independent of θ for all t, the statistic $t(x)$ is said to be an ancillary statistic; a parallel definition can be given for the continuous model. In experiment M^* of Example 1.1, if we set $x = (z, y)$, $z = t(x)$ is an ancillary statistic. Denote by $Pr(y|t(x), \theta)$ the conditional probability given $t(x)$, define the induced sample space by $\Omega' = \{y : t(y) = t(x)\}$, and define the new conditional experiment M'

by $M'(\Omega', \Theta, Pr(\cdot|t(x), \theta))$. Then, the conditionality axiom is expressed by the equality

$$Ev(M, x) = Ev(M', x).$$

However, in general, Basu (1964) showed that situations exist in which the choice of ancillary statistic leads to very different conditional distributions. Fisher emphasized the role of ancillary statistics in the specific circumstance in which information loss is incurred in the process of the maximum-likelihood estimation.

Example 1.2 (Fisher, 1956) Two random variables x and y are independently normally distributed with $E(x) = \cos\theta$, $E(y) = \sin\theta$, and $Var(x) = Var(y) = 1$. Let θ be the unknown parameter. Then, its maximum-likelihood estimate is given by $\hat{\theta} = \tan^{-1}(y/x)$. Set $\hat{l} = \sqrt{x^2 + y^2}$; then, the polar coordinate transformation $(x, y) \rightarrow (\hat{l}, \hat{\theta})$ is one to one and the pair $(\hat{l}, \hat{\theta})$ is sufficient. The distribution of $\hat{l}$ is independent of θ, and hence, $\hat{l}$ is an ancillary statistic. Although the estimate $\hat{\theta}$ alone does not constitute a sufficient statistic, it is sufficient for the conditional distribution of (x, y) given $\hat{l}$. Namely, the lost information from a reduction in the data to $\hat{\theta}$ is recovered by conditioning on $\hat{l}$.

1.4.2 Defining Exogeneity

The exogenous variables play a crucial role for identification and inference on structural equation model (1.1) in Sect. 1.3.1. In that model, the value of the exogenous variable $v(t)$ is assumed to be determined independent of the system (1.1). Such an assumption of exogeneity rather simplifies subsequent inference, whereas it is difficult to find real circumstances to which the assumption applies. The paradigm shift in the 1970s and 1980s of econometrics accompanied the refinement of the concept of exogeneity. When we want to predict future values of endogenous variables or measure the extent of policy effects, we need to address the following questions.

1. In what circumstances are policy variables in a model legitimately regarded as exogenous?

2. In what circumstances is conditional inference plausible given a policy scenario?

3. Under what dynamic conditions we do not need to anticipate feedback from endogenous to exogenous variables?

4. Can we state that policy changes do not accompany shifts in parameters in the behavioral equation of economic agents?

Point (4) is the criticism raised by Lucas (1976) against the conventional economic analyses up to the time. For the simultaneous equation system given in (1.1), $\{v(t)\}$ is defined as strictly exogenous if the sequences $\{\varepsilon(t)\}$ and $\{v(t)\}$ are stochastically independent and the parameter involved in the distribution function of $v(t)$ is independent of the structural parameters A, B and the reduced-form covariance Σ. Statistical inferences on the reduced-form parameters Φ and Σ based on the conditional likelihood given the observations of the strictly exogenous variable do not suffer from information loss. The Cowles approach in the earlier period was mostly based on this definition of exogeneity.

The concept of information amount enables a quantitative characterization of the concepts of sufficiency and ancillarity. Fisher's information amount is a representative one among others. Suppose that the distribution of n random variables $x_1, x_2, \ldots, x_n$ has density $f(x_1, x_2, \ldots, x_n | \theta)$, for which θ is the n_θ-vector parameter. Suppose that density is a smooth function of θ; then, Fisher's information amount (or information matrix) with respect to θ of the random variables is defined by the $n_\theta \times n_\theta$ matrix with the (i, j) component given by

$$I_{ij}(\theta) \equiv E \left\{ \frac{\partial}{\partial \theta_i} \log f(x_1, x_2, \ldots, x_n | \theta) \frac{\partial}{\partial \theta_j} \log f(x_1, x_2, \ldots, x_n | \theta) \right\}, \quad 1 \le i, j \le n_\theta.$$

The quantity expresses the average sensitivity of the likelihood function in response to the parameter change. For standard statistical models, we have the equalities

$$I_{ij}(\theta) = -E \left\{ \frac{\partial^2}{\partial \theta_i \partial \theta_j} \log f(x_1, x_2, \ldots, x_n | \theta) \right\}.$$

Any sufficient statistic has the same information amount as the original data, whereas the information amount of an ancillary statistic is the null matrix.

Regarding the choice of conditioning variables, conditioning on strictly exogenous variables may be an appropriate but not necessary criterion. Fisher introduced the idea of conditional inference in the limited framework of the recovery of lost information of the maximal likelihood estimate by conditioning on an ancillary statistic. Thus, conditioning of an ancillary statistic does not quite fit the predeterminedness of a particular statistic that exogeneity connotes. For large-sample inference, the maximum-likelihood (ML) estimate is generally asymptotically first-order sufficient; however, if measured by the Fisher information amount using its higher-order (second-order) expansion of the distribution, information loss is observed. The loss may be termed the second-order information loss of the ML estimate. However, we can construct a suitable second-order ancillary statistic (namely, a statistic with a second-order asymptotic distributional expansion that is independent of the model parameter), and the lost information of the ML estimate is recovered by conditioning on the constructed second-order ancillary statistic. In other words, the ML estimate is asymptotically higher-order inefficient in general, but can be a conditionally second-order sufficient statistic; see Hosoya (1988).

Hosoya, Tsukuda, and Terui (1989) suggested an approach for recovering the higher-order asymptotic information loss of the limited information ML estimate of a particular over-identified equation in the Cowles Commission model (1.1) through conditioning by suitable ancillary statistics. Specifically, their paper applied the concepts of the curved exponential family of distributions and ancillary statistics to parameter estimations of a single structural equation in a simultaneous equation model and investigated the effect of conditioning on ancillary statistics on the limited information maximum-likelihood (LIML) estimate. The paper also investigated the effect of conditioning on a second-order asymptotic ancillary statistic, i.e., the smallest characteristic root associated with the LIML estimation using an asymptotic expansion of the distribution and the exact distribution. The paper also showed that the smallest root helps provide an intuitively appealing measure of the precision of the LIML estimator.

To define exogeneity, Engle, Hendry, and Richard (EHR) (1983) quoted the concept of S-ancillarity as introduced by Barndorff-Nielsen (1978). Suppose that the conditional probability density function of $x(t) = (u(t), v(t))$ given the past value $X(t - 1) \equiv (x(t - 1), x(t - 2), \dots)$ is factorized as

$$f(x(t)|X(t - 1)|\lambda) = f(u(t)|v(t), X(t - 1)|\lambda_1) f(v(t)|X(t - 1)|\lambda_2), \quad (1.12)$$

and the parameter space Λ of $\lambda = (\lambda_1, \lambda_2)$ has the Cartesian product structure $\Lambda = \Lambda_1 \times \Lambda_2$. Then, EHR defined $v(t)$ to be weakly exogenous with respect to the parameter λ_1. Weak exogeneity is nothing but S-ancillarity for time-series models.

EHR applied the criterion of information loss incurred by conditional inference for classifying the model variables into endogenous and exogenous variables. However, a statistic such as $\hat{l}$ of Example 1.2 satisfies the S-ancillarity condition, whereas they are statistics constructed from endogenous variables and cannot be regarded as variables generated outside the system of interest. The same remarks also apply to the ancillary statistics proposed by Hosoya, Tsukuda, and Terui (1989).

The concept of an ancillary statistic is concerned with a purely statistical inference for recovering the information loss of the maximum-likelihood estimate. In contrast, the dynamic exogeneity of the vector variable $v(t)$ implies that past values of $v(t)$ influence the endogenous variable $u(t)$ but not the other way around regardless of whether or not the model parameter is known. In other words, it is a concept independent of statistical inference and, hence, unrelated to ancillarity. To address this aspect of exogeneity, EHR defined $v(t)$ as strongly exogenous with respect to λ_1 if it is weakly exogenous with respect to λ_1 and, at the same time, if $\{u(t)\}$ does not cause $\{v(t)\}$ in the Granger sense in framework (1.12). Although the EHR paper limits the concept of cause only to the Granger cause in dynamic models, it is not justified. Variables embodying controlled interventions should be well qualified as exogenous; see Neyman's model and the remarks in Sect. 1.1.

To address the Lucas critique, EHR defined $\{v(t)\}$ as super exogenous if it is strongly exogenous and, in addition, if the policy decision accompanying a change of λ_2 leaves λ_1 invariant. The invariance of the parameter values under experimental intervention corresponds to the stability of a noninterventionally observed relation in question.

The original Cowles Commission model focused on how endogenous variables are determined by exterior conditions in terms of comparative statics, whereas the EHR paper is an effort to qualitatively include three quite different aspects in exogeneity. However, the synthesis does not seem successful. Lumping together the different categories of statistical conditional inference, causality, and strict exogeneity into one category does not bring us new insights.

1.4.3 Interpretative Problems

The semantic aspect of Granger causality is certainly not exempt from dispute. Hamilton (1994, pp.306–307) presented the following example. Denote by $P(t)$ and $D(t)$ the price and the dividend of a share on date t, respectively. Suppose that an efficient market provides the equilibrium condition between them

$$P(t) = E_t \sum_{j=1}^{\infty} (1+r)^{-j} D(t+j), \qquad (1.13)$$

where E_t denotes the expectation conditional on the information available up to and including date t. Namely, the conditional expectation of the future flow of dividends determines the price level. However, suppose that the process $\{D(t)\}$ is given as

$$D(t) = c + d + \varepsilon(t) + \delta\varepsilon(t-1) + \eta(t)$$

for mutually independent Gaussian white noise processes $\{\varepsilon(t)\}$ and $\{\eta(t)\}$; then, $\{P(t)\}$ turns out to be a white noise process because $P(t) = (c+d)/r + \delta\varepsilon(t)/(1+r)$ and in no way does $\{D(t)\}$ cause $\{P(t)\}$ in Granger's sense. Hamilton claimed, on the basis of this example, that there are cases for which the true causal relation is not conceptualized by Granger causality.

The problem lies in the interpretation of Eq. (1.13). He took it as a causal relation that the sum of the conditional expectations on the right-hand side determines the left-hand side member. But that interpretation presupposes an extra-model interpretation of causality. That equation only poses a synchronous constraint between $P(t)$ and $E_t D(t+j)$, $j = 1, 2, \cdots$, and what we can assert is, at best, that those quantities are interrelated. The assertion that those conditional expectations cause price $P(t)$ is merely one extra-model belief in this circumstance; moreover, it is of no help if we stand at date $(t-1)$ in predicting those quantities at date t. Because interpretation is concerned with the characteristic of a model that is relevantly termed that of causality or with how causality is modeled, Hamilton's argument seems rather to bring us back to a pre-model stage of conceptualizing causality or to a priori belief of causality.

Simon (1953) and Hicks (1979) emphasized atemporal causality. Although Hicks maintained that causality only makes sense in a model, in his book, he did not present a formal and testable definition of causality. In contrast, Simon considered a system of equations involving a set of variables. He defined a subset of variables X as causing another Y if X is solvable on the basis of a subsystem of equations involving only X and if the solution of Y needs equations involving X. However, placed in a stochastic framework, his definition of causality appears not as compelling. Let ε and η be independent Gaussian random p and q vectors with mean 0 and covariance matrices Σ_{11} and Σ_{22}, respectively. Suppose that observable random vectors X and Y are generated by

$$X = \varepsilon \tag{1.14}$$

$$Y + AX = \eta \tag{1.15}$$

for a $p \times q$ constant matrix A. In this system, X appears to be autonomously determined by (1.14) alone, whereas Y is not and we are tempted to say that X causes Y. However, in terms of the reduced form, X and Y are jointly normally distributed with mean 0 and with covariance matrix

$$\begin{bmatrix} \Sigma_{11} & \Sigma_{11}A^* \\ A\Sigma_{11} & A\Sigma_{11}A^* + \Sigma_{22} \end{bmatrix}$$

and the relation between X and Y is perfectly symmetric, or lacking the asymmetry needed to define causality. Therefore, on the basis of the reduced form, knowledge of realized Y helps predict X and vice versa. The particular structural form given by (1.14) and (1.15) is identifiable and constitutes a testable model, provided that a suitable sample of X and Y is available, say cross-sectionally. However, whether testing that structure in effect tests the one-way causality from X to Y itself depends on an extra-model conception of causality.

Causality is sometimes related to exogeneity, thus shifting the interpretation of causality to that of the latter. Defining variables as exogenous when they are ancillary statistics has its relevance in the statistical inference of parameters in the context of information recovery, but characterizing causal interactions between variables in a system is a different issue. Suppose a stochastic model in which all parameters are specified and known such that inference is not involved. Then, defining exogeneity by ancillarity does not work and, thus, has no role to play in defining causality. However, it still can be of significant interest to understanding the manner in which the variables in the system affect each other.

To conclude the argument, the Sims representation (2.18) to be discussed in Chap. 2 seems to present a pertinent interpretation of Granger's causality in view that the intrinsic innovation component of $v(t - j)$ is temporarily predetermined and is part of the determination of the level of $u(t)$, where the only extra-model assumption we need is that the future does not cause the past.

References

Athanasopoulos, G., & Vahid, F. (2008). VARMA versus VAR for macroeconomic forecasting. *Journal of Business and Economic Statistics, 26*, 237–252.

Barndorff-Nielsen, O. (1978). *Information and Exponential Families: In Statistical Theory.* Chichester: Wiley.

Basu, D. (1964). Recovery of ancillary information. *Sankhya, 21*, 247–256.

Birnbaum, A. (1962). On the foundations of statistical inference (with discussion). *Journal of American Statistical Association, 57*, 269–326.

Cox, D. R. (1958). Some problems connected with statistical inference. *Annals of Mathematical Statistics, 29*, 357–372.

Cox, D. R., & Wermuth, N. (2001). Some statistical aspects of causality. *European Sociological Review, 17*, 65–74.

Engle, R. F., Hendry, D. F., & Richard, J.-F. (1983). Exogeneity. *Econometrica, 51*, 277–304.

Fisher, R. A. (1925). Theory of statistical estimation. *Proceedings of Cambridge Philosophical Society, 22*, 700–725.

Fisher, R.A. (1934). Two new properties of mathematical likelihood, *Proceedings of Royal Society (London), A 144*, 285-307.

Fisher, R. A. (1956). *Statistical methods and scientific inference*. Edinburgh: Oliver and Boyd.

Freedman, D. A. (2010). *Statistical Models and Causal Inference*. Cambridge: Cambridge University Press.

Granger, C. W. J. (1963). Economic process involving feedback. *Information and Control, 6*, 28–48.

Granger, C. W. J., & Hatanaka, M. (1964). *Spectral Analysis of Economic Time Series*. Princeton: Princeton University Press.

Haavelmo, T. (1944). The probability approach in econometrics, *Econometrica, 12*, supplement.

Hamilton, J. D. (1994). *Time Series Analysis*. Princeton: Princeton University Press.

Heckman, J. J. (2000). Causal parameters and policy analysis in economics: A twentieth century retrospective. *The Quarterly Journal of Economics, 115*, 45–97.

Hicks, J. (1979). *Causality in Economics*. Oxford: Basil Blackwell.

Hill, A. B. (1965). The environment and disease: association or causation. *Proceedings of the Royal Society of Medicine, 58*, 295–300.

Holland, P. (1986). Statistics and causal inference (with discussion). *Journal of the American Statistical Association, 81*, 945–970.

Hosoya, Y. (1988). The second-order Fisher information. *Biometrika, 75*, 265–274.

Hosoya, Y., Tsukuda, Y., & Terui, N. (1989). Ancillarity and the limited information maximum-likelihood estimation of a structural equation in a simultaneous equation system. *Econometric Theory, 5*, 384–404.

Johansen, S. (1991). Estimation and hypothesis testing of cointegration vectors in Gaussian vector autoregressive models. *Econometrica, 59*, 1551–1580.

Keynes, J. M. (1939). Professor Tinbergen's method. *The Economic Journal, 49*, 558–570.

Keynes, J. M. (1940). Comment on Tinbergen' s response. *Economic Journal, 50*, 154–156.

Klein, L. R. (1981). *Econometric models as guides for decision-making*. London: The Free Press.

Lucas, R. (1976). Econometric policy evaluation: A critique, the phillips curve and labor markets. *Carnegie Rochester Conference Series on Public Policy, I*, 19–46.

Marschak, J. (1953). Economic measurements for policy and prediction. In William Hood & Tialling Koopmans (Eds.), *Studies in Econometric Method* (pp. 1–26). New York: Wiley.

Marshall, A. (1920). *Principles of Economics* (8th ed.). Londonn: Macmillan and Co. (Reprinted 1930).

Neyman, J. (1923). Sur les applications de la theorie des probabilites aux experiences agricoles: Essai des principes, *Roczniki Nauk Rolniczki, 10*, 1–51, in Polish. English translation by D. Dabrowska & T. Speed (1990), *Statistical Science, 5*, 463–480 (with discussion).

Simon, H. A. (1953). Causal ordering and identifiability. In C. Hood & T. C. Koopmans (Eds.), *Studies in Econometric Method* (pp. 49–74). New York: Wiley.

Sims, C. A. (1980). Macroeconomics and reality. *Econometrica, 48*, 1–47.

Sutton, J. (2000). *Marshall's Tendencies*. Cambridge: The MIT Press.

Tinbergen, J. (1939). *Statistical Testing of Business Cycle Theories*, vols. 1 and 2, League of Nations, Geneva.

Tinbergen, J. (1940). Reply to Keynes. *The Economic Journal, 50*, 141–154.

Wold, H. (1956). Causal inference from observation data. *Journal of the Royal Statistical Society, 119*, 28–60.

Yule, G. U. (1899). An investigation into the causes of changes in pauperism in England, chiefly during the last two intercensal decades. *Journal of the Royal Statistical Society, 62*, 249–295.

Zeisel, H., & Kaye, D. (1997). *Prove It with Figures: Empirical Methods in Law and Litigation*. New York: Springer.

Chapter 2
The Measures of One-Way Effect, Reciprocity, and Association

Abstract To characterize the interdependent structure of a pair of two jointly second-order stationary processes, this chapter introduces the (overall as well as frequency-wise) measures of one-way effect, reciprocity, and association. Section 2.2 defines the Granger and Sims non-causality and establishes their equivalence for a general class of (not necessarily stationary) second-order processes. Sections 2.3 and 2.4 define the overall and frequency-wise one-way effect measures and provide three ways of deriving the frequency-wise measure. One is based on direct canonical factorization of the spectral density matrix. The other two are based on distributed-lag representation and innovation orthogonalization, respectively. Each approach provides a different representation of the same quantity. Section 2.5 introduces the overall and the frequency-wise measures of reciprocity and association.

Keywords Canonical factorization · Frequency-domain representation · Granger non-causality · Measure of association · Measure of one-way effect · Measure of reciprocity · Prediction improvement · Purely reciprocal component process · Sims non-causality

2.1 Prediction and Causality

2.1.1 Statement of the Problem

When we focus on a pair of time series $\{u(t), v(t)\}$ as the subject-matter series, the presence and absence of Granger causality is defined by comparing the prediction precision of the prediction of $u(t)$ using two predictor sets (i) $\{u(s), v(s), s \leq t - 1\}$ and (ii) $\{u(s), s \leq t - 1\}$. The series $\{v(t)\}$ is said to cause or not to cause $\{u(t)\}$ if there is or there is not improvement in prediction accuracy by the predictor set (i) compared with (ii). Absence of Granger causality implies that the addition of information attributable to v up to time $t - 1$ does not improve the prediction accuracy of $u(t)$. This way of defining causality differs from the causality based on intervention, as expounded in Sect. 1.1. The point is that, regardless of how well designed an experiment is, if the knowledge of the cause detected by such an experiment does not

© The Author(s) 2017   21
Y. Hosoya et al., *Characterizing Interdependencies of Multiple Time Series*,
JSS Research Series in Statistics, DOI 10.1007/978-981-10-6436-4_2

help to improve the prediction of the dependent variable in real-life circumstances, the high quality of the experimental design alone is not sufficient for establishing empirical causality.

Granger causality is definable for wide varieties of time-series models, including linear or nonlinear, stationary or non-stationary, and parametric or nonparametric modes. Additionally, a variety of methods exist for measuring prediction improvement. A vast amount of literature exists on estimating and testing Granger causality using time-domain or frequency-domain representations. This book focuses on the mean-square error of the best one-step ahead prediction of second-order processes. To compare the prediction errors of the two prediction methods, it is convenient to use the difference in the log determinants of one-step ahead prediction error covariance matrices. Moreover, for the difference to have harmonic decomposition, namely for the frequency-wise contribution of the difference to be explicitly representable, it is convenient to use as the additional information not the past information $\{v(s), s \leq t - 1\}$ but, instead, its one-way effect components. This one-way effect concept is particularly important in eliminating third-series confounding effects, as discussed in Chap. 3.

The cross-spectrum is a measure of the association between component series that constitutes a multivariate time series. It expresses the covariance between frequency-wise variation elements and is used to quantify short- and long-term time-series interdependence. However, the spectrum itself is not quite fit for characterizing lead or lag dependence between time-series variations. For that purpose, we need knowledge of the canonical factor of the spectral density matrix of the subject-matter time series and the allied prediction theory.

2.1.2 Terminology and Notations

The following notations and symbols are used throughout. The sets of all integers, nonnegative integers, non-positive integers, and positive integers are denoted, respectively, by $\mathbb{Z}$, $\mathbb{Z}^{0+}$, $\mathbb{Z}^{0-}$, and $\mathbb{Z}^{+}$. For a set of random variables $\{w_i, i \in \mathbb{A}\}$ with finite second moment, $H\{w_i, i \in \mathbb{A}\}$ implies the closure in the mean square of the linear hull of $\{w_i, i \in \mathbb{A}\}$ in the Hilbert space of all random variables with finite second moment. For a p-vector process $x(t)$ with finite covariance matrix and for a set of integers $\mathbb{S}$, $H\{x(t), t \in \mathbb{S}\}$ implies $H\{x_i(t), t \in \mathbb{S}, i = 1, \cdots, p\}$. For a notational economy, $H\{x(t_1 - j), y(t_2 - j); \ j \in \mathbb{Z}^{0+}\}$ is written simply as $H\{x(t_1), y(t_2)\}$, and $H\{x(j); j \in \mathbb{Z}\}$ is written as $H\{x(\infty)\}$. See Appendix A.1 for a brief introduction of Hilbert space.

We identify an information set of variables with the Hilbert subspace generated by the set of variables. Thus, we identify the linear prediction of a variable using the information set of predictor variables with the orthogonal projection of the predicted variable onto the Hilbert space generated by the predictor variables, where the projection is the best linear predictor and the accompanying prediction error is the residual (or perpendicular) of the projection. Namely, we translate the linear prediction problems to those of projections onto Hilbert subspaces.

This book primarily follows the standard notations and the basic framework of the theory; however, the one-way effect concept requires projections of random vectors onto special Hilbert subspaces. Let $\{u(t); t \in \mathbb{Z}\}$ and $\{v(t); t \in \mathbb{Z}\}$ ($\mathbb{Z}$: the set of all integers) be, respectively, real p_1- and p_2-dimensional second-order processes with mean 0 defined on a common probability space; a stochastic process with finite covariances is termed a second-order process. Denote by H the Hilbert subspace spanned by all the component random variables of $\{u(t), v(t); t \in \mathbb{Z}\}$ in the Hilbert space of all random variables with finite second moment. In this book, the projection of a random vector $z = \{z_j; j = 1, \ldots, r\}$ onto $H\{\cdot\}$, a subspace of H, implies component-wise projection. Namely, if $\tilde{z}_j$ is the projection of z_j onto $H\{\cdot\}$, then the projection of z onto $H\{\cdot\}$ implies the vector $\tilde{z}$, whose jth component is $\tilde{z}_j$. If each component z_j of a vector z belongs to $H\{\cdot\}$, the random vector z is said to belong to $H\{\cdot\}$. The difference $z - \tilde{z}$ is termed the residual of the projection of z onto $H\{\cdot\}$. If each component z_j is orthogonal to $H\{\cdot\}$, z is said to be orthogonal to $H\{\cdot\}$ and denoted as $z \perp H\{\cdot\}$. Two subspaces H_1 and H_2, which are orthogonal to each other, are denoted as $H_1 \perp H_2$. The orthogonal complement of $H\{\cdot\}$ in $H \equiv H\{u(\infty), v(\infty)\}$ is denoted as $H\{\cdot\}^\perp$. The subscripts and prime attached to u and v denote the subspace to which a concerned projection is related, and the upper bar denotes the projection. For example, for a second-order stationary process $\{u(t), v(t)\}$, $u_{-1,.}(t)$, $u_{-1,-1}(t)$, and $u_{-1,0}(t)$ are the residuals of the projection of $u(t)$ onto $H\{u(t-1)\}$, $H\{u(t-1), v(t-1)\}$, and $H\{u(t-1), v(t)\}$, respectively, and $\bar{u}_{-1,.}(t)$, $\bar{u}_{-1,-1}(t)$, and $\bar{u}_{-1,0}(t)$ are the corresponding projections. The prime denotes that $u(t)$ and $v(t)$ are projected subspaces of $H\{u(\infty), v_{0,-1}(\infty)\}$ and $H\{u_{-1,0}(\infty), v(\infty)\}$, respectively, such that $u'_{-1,-1}(t)$ and $u'_{.,\infty}(t)$ are the residuals of the projection of $u(t)$ onto $H\{u(t-1), v_{0,-1}(t-1)\}$, and $H\{v_{0,-1}(\infty)\}$, whereas $v'_{-1,-1}(t)$ and $v'_{\infty,.}(t)$ are the projections of $v(t)$ onto $H\{u_{-1,0}(t-1), v(t-1)\}$ and $H\{u_{-1,0}(\infty)\}$.

To quantify the contribution of the process $\{v(t)\}$ to the improvement of the one-step ahead prediction of $\{u(t)\}$, Geweke (1982) introduced the log-ratio

$$F_{v \to u} = \log[\det\{Cov(u_{-1,.}(t))\}/\det\{Cov(u_{-1,-1}(t))\}], \tag{2.1}$$

and termed it the measure of linear feedback. To represent the frequency-wise contribution of the process $\{v(t)\}$ to $\{u(t)\}$ when they jointly constitute an autoregressive process, he proposed a nonnegative function $F_{v \to u}(\lambda)$, showing under certain conditions that $F_{v \to u}$ can be represented as the integral of $F_{v \to u}(\lambda)$ over the frequency domain. In contrast, Akaike (1968) introduced the measure termed relative power contribution (RPC), which quantifies the frequency-wise contribution of a component noise process to an observation series in a multivariate autoregressive process. Through a suitable extension, his RPC measure is related to the one-way effect measure, as is shown in Sect. 2.5. [See Granger (1969) and Pierce (1979) for other proposals of measurements of interdependency.]

Because $\{v_{0,-1}(t); t \in \mathbb{Z}\}$ can be regarded as the proper innovation process contained in $\{v(t)\}$, Sect. 2.3 of this chapter proposes as the (overall) measure of the one-way effect of the process $\{v(t)\}$ to $\{u(t)\}$ the log-ratio

$$M_{v \to u} = \log[\det\{Cov(u_{-1,.}(t))\}/\det\{Cov(u'_{-1,-1}(t))\}].$$

In contrast to $F_{v \to u}$, the measure $M_{v \to u}$ has a natural decomposition into a frequency-wise measure $M_{v \to u}(\lambda)$, which is introduced in (2.16) such that

$$M_{v \to u} = \frac{1}{2\pi} \int_{-\pi}^{\pi} M_{v \to u}(\lambda) d\lambda$$

in general circumstances (Theorem 2.2). Section 2.4 constructs $M_{v \to u}(\lambda)$ by two other approaches and, thus, provides alternative representations. Theorem 2.3 shows that one of the constructed measures $\tilde{M}_{v \to u}(\lambda)$ is equal to $M_{v \to u}(\lambda)$. Moreover, in that section, $\tilde{M}_{v \to u}(\lambda)$ is shown to be equal to Geweke's $F_{v \to u}(\lambda)$ for the autoregressive process; thus, $M_{v \to u}(\lambda)$ is an extension of the latter.

Section 2.5 introduces the measures of association and reciprocity in the frequency domain and shows that the measure of association is decomposed into the sum of the measures of the one-way effect and of reciprocity [Theorem 2.5]. Section 2.5 also provides a sufficient condition for the corresponding overall measures to be equal to the Gel'fand–Yaglom measure (2.34) and the Geweke measure (2.1), respectively [Theorem 2.6]. Remark 2.3 discusses the relationship of the one-way effect measure with Akaike's RPC measure. Section 2.6 provides two examples for illustration purposes. One example is the case in which a process is not invertible but $M_{v \to u}(\lambda)$ is measured. The other example illustrates a situation in which $F_{v \to u} > M_{v \to u}$.

Regarding further notations used in this book, $\{x(t)\}$ denotes the process $\{x(t); t \in \mathbb{Z}\}$ unless otherwise specified. For a specified partition of a $(p_1 + p_2) \times (p_1 + p_2)$ matrix A

$$A = \begin{bmatrix} A_{11} & A_{12} \\ A_{21} & A_{22} \end{bmatrix}$$

always implies that A_{11} is a $p_1 \times p_1$ submatrix. Two random variables that are equal with probability 1 are identified such that the a.e. notation is omitted. For a random vector x or for a pair of random vectors x and y, $Cov(x)$ and $Cov(x, y)$ indicate the variance-covariance matrix of x and $vec(x, y)$, respectively. The identity matrix of order p is denoted by I_p. A^* indicates the conjugate transpose if A is a complex matrix and the simple transpose if A is a real matrix. Sometimes, A' is also used to denote the transpose of a real matrix A. The trace of a square matrix C is denoted by $tr\, C$, and the determinant is denoted by $\det C$. The sum of two vector subspaces H_1 and H_2 is denoted by $H_1 \oplus H_2$. The symbol $\equiv$ is used for definitions.

2.2 Defining Non-causality

This section expounds on the concepts of Granger's non-causality and Sims's counterpart explicitly for a pair of vector-valued second-order processes [See Granger (1963, 69), Sims (1972), Hosoya (1977), and Florens and Mouchart (1982) for related literature.]. Let $\{u(t)\}$ and $\{v(t)\}$ be, respectively, real p_1- and p_2-vector (not necessarily stationary) stochastic processes with mean 0 and finite second-order moments defined in a common probability space. The Granger condition for non-causality in terms of the mean-square prediction error is defined as follows. [Granger throughout attributed the idea to N. Wiener.]

Definition 2.1   The process $\{v(t)\}$ does not cause $\{u(t)\}$ if the projection of $u(t)$ onto $H\{u(t-1), v(t-1)\}$ belongs to $H\{u(t-1)\}$ for all $t \in \mathbb{Z}$.

Lemma 2.1   *This Granger condition is equivalent to the relationship*

$$u_{-1,.}(t) \perp H\{u(t-1), v(t-1)\}, \quad t \in \mathbb{Z},$$

where $u_{-1,.}(t)$ denotes the residual of the projection of $u(t)$ onto $H\{u(t-1)\}$.

Proof   Suppose that $u_{-1,.}(t) \perp H\{u(t-1), v(t-1)\}$. Because the projection of $u_{-1,.}(t)$ onto $H\{u(t-1), v(t-1)\}$ is 0 and because the projection of $\bar{u}_{-1,.}(t)$ onto $H\{u(t-1), v(t-1)\}$ is $\bar{u}_{-1,.}(t)$ itself, the projection of $u(t)$ onto $H\{u(t-1), v(t-1)\}$ is equal to $\bar{u}_{-1,.}(t)$, which belongs to $H\{u(t-1)\}$. Hence, the Granger condition follows. To prove the reverse implication, suppose that $\bar{u}_{-1,-1}(t) = \bar{u}_{-1,.}(t)$, whence we have $u_{-1,-1}(t) = u_{-1,.}(t)$. Because $u_{-1,-1}(t) \perp H\{u(t-1), v(t-1)\}$, the assertion $u_{-1,.}(t) \perp H\{u(t-1), v(t-1)\}$ follows. $\square$

Denote by $\bar{v}_{0,.}(t)$ the projection of $v(t)$ onto $H\{u(t)\}$, and set $v_{0,.}(t) = v(t) - \bar{v}_{0,.}(t)$. The decomposition $v(t) = \bar{v}_{0,.}(t) + v_{0,.}(t), t \in \mathbb{Z}$ is the Sims distributed-lag representation of $v(t)$ by the process $\{u(t)\}$, where we have $\bar{v}_{0,.}(t) \in H\{u(t)\}$ and $v_{0,.}(t) \in H\{u(t)\}^{\perp}$. See (1.4) for the autoregressive distributed-lag representation.

Definition 2.2   The Sims condition for $\{v(t)\}$ not causing $\{u(t)\}$ implies that $v_{0,.}(t)$ is orthogonal to $H\{u(\infty)\}$ for all t, where $v_{0,.}(t)$ is the error term in the Sims distributed-lag representation of $v(t)$ by $u(t-j), j \in \mathbb{Z}^{0+}$, and $H\{u(\infty)\}$ denotes the completion of the linear hull of the set $\{u(t), t \in \mathbb{Z}\}$.

Theorem 2.1   *The Granger condition is equivalent to the Sims condition.*

Proof   First, we show that the Sims condition follows from the Granger condition. Because $v_{0,.}(t) \perp H\{u(t)\}$, the problem is to show that $u(t+p) \perp v_{0,.}(t)$ for any $p \geq 1$ and $t \in \mathbb{Z}$. Set $h(t+p-1) = \bar{u}_{-1,.}(t+p)$, and set $\varepsilon(t+p) \equiv u_{-1,.}(t+p)$. For $i = 2, \ldots, p$, let $h(t+p-i)$ be the projection of $h(t+p-i+1)$ onto $H\{u(t+p-i)\}$ and set the residual as $\varepsilon(t+p-i+1) = h(t+p-i+1) - h(t+p-i)$. Then, by iterative projection we have

$$u(t+p) = h(t+p-1) + \varepsilon(t+p)$$
$$= h(t+p-2) + \{\varepsilon(t+p) + \varepsilon(t+p-1)\}$$
$$= h(t) + \sum_{i=0}^{p-1} \varepsilon(t+p-i) \equiv h(t) + \xi(t+p).$$

In view of Lemma 2.1, the Granger non-causality implies

$$\varepsilon(t+p) \perp H\{u(t+p-1), v(t+p-1)\}.$$

Denote by $J\{u(t+p-1)\}$ the vector space spanned singly by $u(t+p-1)$. Then, we have $h(t+p-1) \equiv \bar{u}_{-1,.}(t+p) \in J\{u(t+p-1)\} \oplus H\{u(t+p-2)\}$; namely, $h(t+p-1)$ is expressed as

$$h(t+p-1) = c_1 u(t+p-1) + l(t+p-2), \tag{2.2}$$

where c_1 is a constant, $p_1 \times p_1$ is a matrix, and $l(t+p-2) \in H\{u(t+p-2)\}$. By projecting $h(t+p-1)$ onto $H\{u(t+p-2)\}$, we have

$$h(t+p-1) = h(t+p-2) + \varepsilon(t+p-1)$$

where $\varepsilon(t+p-1)$ is the projection residuals of $c_1 u(t+p-1)$; hence, it follows from the Granger condition and (2.2) that

$$\varepsilon(t+p-1) \perp H\{u(t+p-2), v(t+p-2)\}.$$

By repeating a similar argument, we can conclude that $h(t) \in H\{u(t)\}$ and $\xi(t+p) \perp H\{u(t), v(t)\}$. Because then $v_{0,.}(t) \in H\{u(t), v(t)\}$, we have the orthogonality $\xi(t+p) \perp v_{0,.}(t)$. Then, because $h(t) \in H\{u(t)\}$ and $v_{0,.}(t) \perp H\{u(t)\}$, we have $v_{0,.}(t) \perp h(t)$. Consequently, the relation $v_{0,.}(t) \perp u(t+p)$ follows.

To prove the reverse implication, suppose that the distributed-lag representation of $v(t)$ satisfies the Sims condition; namely, suppose that the subspace $H\{v(t-1)\}$ is included in the vector sum $H\{u(t-1)\} \oplus H\{u(\infty)\}^\perp$. If $g \in H\{v(t-1)\}$, it is a limit in the mean-square distance of the finite sum $\sum_{j=0}^{p} c_j v(t-j-1) \in H\{v(t-1)\} \subset H\{u(t-1)\} \oplus H\{u(\infty)\}^\perp$. Hence, we have $H\{v(t-1), v(t-1)\} \subset H\{u(t-1)\} \oplus H\{u(\infty)\}^\perp$. Because $u_{-1,.}(t) \in H\{u(\infty)\}$, we have $u_{-1,.}(t) \perp H\{u(\infty)\}^\perp$. Supposing that $h(t)$ is an element of $H\{v(t-1)\}$, set $h(t) = l(t) + \lambda(t)$ where $l(t) \in H\{u(t-1)\}$ and $\lambda(t) \in H\{u(\infty)\}^\perp$. The relations $u_{-1,.}(t) \perp l(t)$ and $u_{-1,.}(t) \perp \lambda(t)$ hold, but they imply that $u_{-1,.}(t) \perp h(t)$. Because $h(t)$ is arbitrary, $u_{-1,.}(t) \perp H\{v(t-1)\}$. Thus, $u_{-1,.}(t)$ is orthogonal to $H\{u(t-1), v(t-1)\}$. It follows then from Lemma 2.1 that the Granger condition holds for the process $\{u(t), v(t)\}$. $\qquad\square$

Corollary 2.1 $\bar{u}'_{-1,-1}(t) \in H\{u(t-1)\}$ *if and only if* $v_{0,-1}(t) \in H(u(\infty))^\perp$.

Proof The Granger condition that $\{v_{0,-1}(t)\}$ does not cause $\{u(t)\}$ is the left-hand side condition, whereas the Sims condition is given by $v_{0,-1}(t) \in H\{u(t)\} \oplus H(u(\infty))^{\perp}$. However, because the projection of $v_{0,-1}(t)$ onto $H\{u(t)\}$ is 0, the corollary follows. $\square$

The next corollary asserts that $\{v(t)\}$ does not cause $\{u(t)\}$ if and only if $\{v_{0,-1}(t)\}$ does not cause $\{u(t)\}$ under the assumption of pure non-determinism.

Corollary 2.2 *Suppose that $\{v(t)\}$ is purely non-deterministic, namely $\bigcap_{j=0}^{\infty} H\{v(t-j)\} = \{0\}$. Then, $\bar{u}_{-1,-1}(t)$ belongs to $H\{u(t-1)\}$ if and only if $\bar{u}'_{-1,-1}(t)$ belongs to $H\{u(t-1)\}$.*

Proof Because $H\{u(t-1), v_{0,-1}(t-1)\} \subset H\{u(t-1), v(t-1)\}$, the necessity is evident.

The sufficiency is shown as follows. Corollary 2.1 implies that $v_{0,-1}(t) \in H^{\perp}\{u(\infty)\}$ and, by definition, $\bar{v}_{0,-1}(t) \in \{u(t), v(t-1)\}$. Therefore, we have

$$v(t) = \bar{v}_{0,-1}(t) + v_{0,-1}(t) \in H\{u(t), v(t-1)\} \oplus H^{\perp}\{u(\infty)\}. \tag{2.3}$$

Hence

$$H\{u(t), v(t)\} \oplus H^{\perp}\{u(\infty)\} = H\{u(t), v(t-1)\} \oplus H^{\perp}\{u(\infty)\}.$$

Repeating the same argument, we have

$$H\{u(t), v(t)\} \oplus H^{\perp}\{u(\infty)\} = H\{u(t), v(t-j)\} \oplus H^{\perp}\{u(\infty)\} \tag{2.4}$$

for any positive integer j. Thanks to the assumption, $\bigcap_{j=0}^{\infty} H\{v(t-j)\} = \{0\}$. It then follows from (2.4) that

$$H\{u(t), v(t-1)\} \oplus H^{\perp}\{u(\infty)\} \subset H\{u(t)\} \oplus H^{\perp}\{u(\infty)\},$$

whence in view of (2.3) we have $v(t) \in H\{u(t)\} \oplus H^{\perp}\{u(\infty)\}$. Accordingly, Corollary 2.2 follows from Corollary 2.1. $\square$

2.3 The One-Way Effect Measure

The one-way effect from one process to another is quantitatively characterized using the prediction theory of stationary processes. Let $\{u(t), v(t), t \in \mathbb{Z}\}$ be a zero mean jointly second-order non-deterministic full-rank stationary process where the $u(t)$ and $v(t)$ are real p_1 and p_2 vectors, respectively. Suppose also that the process has the $p \times p$ spectral density matrix

$$h(\lambda) = \begin{bmatrix} h_{11}(\lambda) & h_{12}(\lambda) \\ h_{21}(\lambda) & h_{22}(\lambda) \end{bmatrix}, \quad -\pi < \lambda \le \pi,$$

where $(p = p_1 + p_2)$ and $h_{11}(\lambda)$ is the $p_1 \times p_1$ spectral density of $\{u(t)\}$. Namely, the second-order stationarity implies that the covariance matrix

$$E \begin{bmatrix} u(t+s) \\ v(t+s) \end{bmatrix} [u(t)^*, v(t)^*] = V(s)$$

does not depend on t, for any $t, s \in \mathbb{Z}$. The matrix $V(s)$ is termed serial covariance matrix, and we assume it is representable as

$$V(s) = \int_{-\pi}^{\pi} h(\lambda) e^{i\lambda s} d\lambda, \quad s \in \mathbb{Z}$$

by means of an Hermite matrix-valued function $h(\lambda)$. The matrix $h(\lambda)$ is termed the spectral density matrix of the process $\{u(t), v(t)\}$.

Assumption 2.1 The spectral density matrix of the process $\{u(t), v(t)\}$ satisfies the Szegö condition

$$\int_{-\pi}^{\pi} \log \det h(\lambda) d\lambda > -\infty. \tag{2.5}$$

A process is said to have maximal rank if it has a full-rank spectral density matrix $a.e.$ on $(-\pi, \pi]$. The Szegö condition is necessary and sufficient for $\{u(t), v(t)\}$ being purely non-deterministic and $\det h(\lambda) > 0 \, a.e.$ with respect to the Lebesgue measure on the torus, see Rozanov (1967, p. 73). The condition is assumed throughout the sequel. Under the condition (2.5), $h(\lambda)$ is known to have a factorization such that

$$h(\lambda) = \frac{1}{2\pi} \Gamma(e^{-i\lambda}) \Gamma(e^{-i\lambda})^*, \tag{2.6}$$

where $\Gamma(e^{-i\lambda})$ is the boundary value $\lim_{\mu \to 1-} \Gamma(\mu e^{-i\lambda})$ of a $p \times p$ matrix-valued function $\Gamma(z)$, which is analytic, and $\det \Gamma(z)$ has no zeros inside the unit disk $\{z : |z| < 1\}$ of the complex plane. Let $\Gamma_1(z)$ be another such function satisfying those same conditions of $\Gamma(z)$ including the boundary condition (2.6). If there is no $\Gamma_1(z)$ such that $\Gamma_1(0)\Gamma_1(0)^* - \Gamma(0)\Gamma(0)^*$ is positive definite, the factor $\Gamma(z)$ is said maximal.

Such a factorization (2.6) by means of a maximal $\Gamma(z)$ is said to be a canonical factorization in the sequel. Denote the covariance matrix of the one-step ahead prediction error of the process $\{u(t), v(t)\}$ by

$$\Sigma = \begin{bmatrix} \Sigma_{11} & \Sigma_{12} \\ \Sigma_{21} & \Sigma_{22} \end{bmatrix}$$

where $\Sigma = Cov\{u_{-1,-1}(t), v_{-1,-1}(t)\}$. Thanks to the Szegö condition, Σ is positive definite. For those discussions and the next lemma, see Rozanov (1967, pp. 71–77).

Lemma 2.2 $\Gamma(z)$ *is maximal if and only if*

$$\det\{\Gamma(0)\Gamma(0)^*\} = \det \Sigma = (2\pi)^p \exp\{\frac{1}{2\pi} \int_{-\pi}^{\pi} \log \det h(\lambda)d\lambda\}. \qquad (2.7)$$

We define $M_{v\rightarrow u}$ as the (overall) measure of one-way effect from v to u by

$$M_{v\rightarrow u} = \log[\det\{Cov(u_{-1,.}(t))\}/\det\{Cov(u'_{-1,-1}(t))\}]. \qquad (2.8)$$

Because $\det Cov(u_{-1,-1}(t)) \leq \det Cov(u'_{-1,-1}(t))$, we have the next lemma.

Lemma 2.3 *Let $F_{v\rightarrow u}$ be the Geweke measure given by*

$$F_{v\rightarrow u} \equiv \log[\det\{Cov(u_{-1,.}(t))\}/\det\{Cov(u_{-1,-1}(t))\}].$$

We have $0 \leq M_{v\rightarrow u} \leq F_{v\rightarrow u}$. The equality $M_{v\rightarrow u} = F_{v\rightarrow u}$ holds if and only if $\bar{u}_{-1,-1}(t)$ belongs to $H\{u(t-1), v_{0,-1}(t-1)\}$.

Example 2.2 of Sect. 2.6 provides the case in which $M_{v\rightarrow u} < F_{v\rightarrow u}$. As is seen in the construction of $v_{0,-1}(t)$, $M_{v\rightarrow u}$ quantifies the extent of the proper contribution by the component of $\{v(t)\}$, which does not contain the feedback effect from $\{u(t)\}$.

The following three lemmas are preliminary to the construction of the frequency-wise measure $M_{v\rightarrow u}(\lambda)$.

Lemma 2.4 *The process $\{u(t), v_{0,-1}(t)\}$ has the maximal rank and is purely non-deterministic. Additionally, we have*

$$v_{0,-1}(t) = v_{-1,-1}(t) - \Sigma_{21}\Sigma_{11}^{-1}u_{-1,-1}(t). \qquad (2.9)$$

Moreover, $\{v_{0,-1}(t)\}$ is a white noise process with covariance matrix $\Sigma_{22} - \Sigma_{21}\Sigma_{11}^{-1}\Sigma_{12}$.

Proof In the decomposition

$$\begin{aligned}
v(t) &= \bar{v}_{-1,-1}(t) + v_{-1,-1}(t) \\
&= \{v_{-1,-1}(t) - \Sigma_{21}\Sigma_{11}^{-1}u_{-1,-1}(t)\} + \{\Sigma_{21}\Sigma_{11}^{-1}u_{-1,-1}(t) + \bar{v}_{-1,-1}(t)\},
\end{aligned}$$

we have $\{v_{-1,-1}(t) - \Sigma_{21}\Sigma_{11}^{-1}u_{-1,-1}(t)\} \perp H\{u_{-1,-1}(t), u(t-1), v(t-1)\}$ and $\{\Sigma_{21}\Sigma_{11}^{-1}u_{-1,-1}(t) + \bar{v}_{-1,-1}(t)\} \in H\{u(t), v(t-1)\}$. By the uniqueness of the projection residual, we have

$$v_{0,-1}(t) = v_{-1,-1}(t) - \Sigma_{21}\Sigma_{11}^{-1}u_{-1,-1}(t).$$

Next, suppose that a scalar-valued process $\{s(t)\}$ is given by

$$
\begin{aligned}
s(t) &= \alpha^* u(t) + \beta^* v_{0,-1}(t) \\
&= \alpha^* \bar{u}_{-1,-1}(t) + \{\gamma^* u_{-1,-1}(t) + \beta^* v_{-1,-1}(t)\},
\end{aligned}
$$

where α, β are real p_1 and p_2 vectors, respectively, not all zero, and $\gamma^* \equiv \alpha^* - \beta^* \Sigma_{21} \Sigma_{11}^{-1}$. That the process $\{u(t), v_{0,-1}(t)\}$ has maximal rank is proven if the spectral density $h_s(\lambda)$ of the process $\{s(t)\}$ is positive $a.e.$. The first sum is orthogonal to the last two sums, whereas the sum of those last two constitutes a white noise with positive variance because γ and β cannot be simultaneously zero unless α and β are all simultaneously zero. Therefore, the spectral density $h_s(\lambda)$ is the sum of the spectral density $\alpha^* \bar{u}_{-1,-1}(t)$ and the spectral density of the sum of the last two terms, which is a positive constant. Therefore, $h_s(\lambda) > 0$. The proposition that the process $\{u(t), v_{0,-1}(t)\}$ is purely non-deterministic follows from the relations

$$
\bigcap_{j=0}^{\infty} H\{u(t-j), v_{0,-1}(t-j)\} \subset \bigcap_{j=0}^{\infty} H\{u(t-j), v(t-j)\} = \{0\}.
$$

Because $\{u_{-1,-1}(t), v_{-1,-1}(t)\}$ is a white noise process, $\{v_{0,-1}(t)\}$ is also a white noise process in view of (2.9).   □

Let $\tilde{h}$ and $\check{h}$ be the spectral densities of the joint processes $\{u(t), v_{0,-1}(t)\}$ and $\{u_{-1,0}(t), v(t)\}$, and denote the partitioned matrices by

$$
\tilde{h}(\lambda) = \begin{pmatrix} \tilde{h}_{11}(\lambda) & \tilde{h}_{12}(\lambda) \\ \tilde{h}_{21}(\lambda) & \tilde{h}_{22}(\lambda) \end{pmatrix}; \quad \check{h}(\lambda) = \begin{pmatrix} \check{h}_{11}(\lambda) & \check{h}_{12}(\lambda) \\ \check{h}_{21}(\lambda) & \check{h}_{22}(\lambda) \end{pmatrix}. \tag{2.10}
$$

In view of the construction, it is evident that $\tilde{h}_{11}(\lambda) = h_{11}(\lambda)$, $\tilde{h}_{22}(\lambda) = (\Sigma_{22} - \Sigma_{21} \Sigma_{11}^{-1} \Sigma_{12})/(2\pi) \equiv \Sigma_{22:1}/(2\pi)$, $\tilde{h}_{21}(\lambda) = \tilde{h}_{12}(\lambda)^*$; $\check{h}_{11}(\lambda) = (\Sigma_{11} - \Sigma_{12} \Sigma_{22}^{-1} \Sigma_{21})/(2\pi) \equiv \Sigma_{11:2}/(2\pi)$ and $\check{h}_{22}(\lambda) = h_{22}(\lambda)$, $\check{h}_{21}(\lambda) = \check{h}_{12}(\lambda)^*$. Expressions for $\tilde{h}_{12}(\lambda)$ and $\check{h}_{12}(\lambda)$ are given in Lemma 2.5, where we note that the inverse $\Gamma(e^{-i\lambda})^{-1}$ exists $a.e.$ because h has maximal rank.

Lemma 2.5 $\tilde{h}_{12}(\lambda)$ and $\check{h}_{12}(\lambda)$ are represented as:

$$
\tilde{h}_{12}(\lambda) = h_{1\cdot} \Gamma(e^{-i\lambda})^{-1*} \Gamma(0)^* (-\Sigma_{21} \Sigma_{11}^{-1}, I_{p_2})^*, \tag{2.11}
$$

$$
\check{h}_{12}(\lambda) = (I_{p_1}, -\Sigma_{12} \Sigma_{22}^{-1}) \Gamma(0) \Gamma(e^{-i\lambda})^{-1} h_{\cdot 2}(\lambda), \tag{2.12}
$$

where $h_{1\cdot}(\lambda)$ is a $p_1 \times (p_1 + p_2)$ matrix that consists of the first p_1 rows of $h(\lambda)$ and $h_{\cdot 2}(\lambda)$ denotes the $(p_1 + p_2) \times p_2$ matrix consisting of the last p_2 columns of $h(\lambda)$.

Proof Set $A \equiv (-\Sigma_{21}\Sigma_{11}^{-1}, I_{p_2})$. Denote by $\Phi_u(d\lambda)$ and $\Phi_v(d\lambda)$ the orthogonal-increment random spectral measures of the process $\{u(t)\}$ and $\{v(t)\}$; namely, we have

$$u(t) = \int_{-\pi}^{\pi} e^{i\lambda t} \Phi_u(d\lambda) \quad \text{and} \quad v(t) = \int_{-\pi}^{\pi} e^{i\lambda t} \Phi_v(d\lambda).$$

Because $\Gamma(e^{-i\lambda})$ is invertible, the innovation process for $\{u(t), v(t)\}$ is given by

$$\begin{bmatrix} u_{-1,-1}(t) \\ v_{-1,-1}(t) \end{bmatrix} = \int_{-\pi}^{\pi} e^{i\lambda t} \Gamma(0)\Gamma(e^{-i\lambda})^{-1} \begin{bmatrix} \Phi_u(d\lambda) \\ \Phi_v(d\lambda) \end{bmatrix};$$

whence the one-way effect component of $v(t)$ is represented by

$$\begin{aligned} v_{0,-1}(t) &= v_{-1,-1}(t) - \Sigma_{21}\Sigma_{11}^{-1} u_{-1,-1}(t) \\ &= \int_{-\pi}^{\pi} e^{i\lambda t} A\Gamma(0)\Gamma(e^{-i\lambda})^{-1} \begin{bmatrix} \Phi_u(d\lambda) \\ \Phi_v(d\lambda) \end{bmatrix}. \end{aligned}$$

Then, the covariance matrix $Cov(u(t), v_{0,-1}(s))$ is expressed in the Fourier transform as

$$\begin{aligned} Cov\{u(t), v_{0,-1}(s)\} &= \int_{-\pi}^{\pi} e^{i\lambda(t-s)} E\{\Phi_u(d\lambda) \begin{bmatrix} \Phi_u(d\lambda) \\ \Phi_v(d\lambda) \end{bmatrix}^* \}\Gamma(e^{-i\lambda})^{-1*}\Gamma(0)^* A^* \\ &= \int_{-\pi}^{\pi} e^{i\lambda(t-s)} \{h_{11}(\lambda), h_{12}(\lambda)\}\Gamma(e^{-i\lambda})^{-1*}\Gamma(0)^* A^* d\lambda, \end{aligned}$$

(2.13)

where $\{h_{11}(\lambda), h_{12}(\lambda)\}$ is the $p_1 \times (p_1 + p_2)$ upper submatrix of $h(\lambda)$. The expression (2.11) follows from (2.13). The expression (2.12) is derived from a parallel argument. $\square$

Set $h_{11:2}(\lambda) = h_{11}(\lambda) - 2\pi \tilde{h}_{12}(\lambda)\Sigma_{22:1}^{-1}\tilde{h}_{21}(\lambda)$ where $\Sigma_{22:1} = \Sigma_{22} - \Sigma_{21}\Sigma_{11}^{-1}\Sigma_{12}$; then we have:

Lemma 2.6 *The covariance matrix $Cov(u'_{-1,-1}(t))$ has the following decomposability property in the frequency domain:*

$$\det Cov(u'_{-1,-1}(t)) = (2\pi)^{p_1} \exp\left\{ \frac{1}{2\pi} \int_{-\pi}^{\pi} \log \det h_{11:2}(\lambda)d\lambda \right\}. \quad (2.14)$$

Proof Because $u'_{-1,-1}(t) \perp v_{0,-1}(t)$, we have

$$\det Cov\{u'_{-1,-1}(t), v_{0,-1}(t)\} = \det Cov\{u'_{-1,-1}(t)\} \det Cov\{v_{0,-1}(t)\}.$$

Then, in view of Lemma 2.4, we have

$$
\begin{aligned}
&\log \det Cov\{u'_{-1,-1}(t), v_{0,-1}(t)\} \\
&= \log \left[(2\pi)^{p_1+p_2} \exp \left\{ \frac{1}{2\pi} \int_{-\pi}^{\pi} \log \det \tilde{h}(\lambda) d\lambda \right\} \right] \\
&= \frac{1}{2\pi} \int_{-\pi}^{\pi} \left[\log \det \tilde{h}_{22}(\lambda) + \log \det \left\{ h_{11}(\lambda) - \tilde{h}_{12}(\lambda)\tilde{h}_{22}^{-1}(\lambda)\tilde{h}_{21}(\lambda) \right\} \right] d\lambda \\
&\quad + (p_1 + p_2) \log(2\pi).
\end{aligned}
\tag{2.15}
$$

Because $\{v_{0,-1}(t)\}$ is a white noise process with the spectral density $\tilde{h}_{22}(\lambda) = \Sigma_{22:1}/(2\pi)$ such that $\det Cov\{v_{0,-1}(t)\} = \det \Sigma_{22:1}$. The relation (2.14) follows from (2.15). $\qquad\square$

Using the decomposition of Lemma 2.6, define the measure $M_{v \to u}(\lambda)$ of the one-way effect from v to u at frequency λ by

$$
M_{v \to u}(\lambda) = \log\{\det h_{11}(\lambda)/\det h_{11:2}(\lambda)\}, \quad -\pi < \lambda \le \pi.
\tag{2.16}
$$

Note that the expression $h_{11:2}(\lambda)$ in the denominator in (2.16) denotes the spectral density matrix of the process $\{u'_{,\infty}(t)\}$. It is evident that $M_{v \to u}(\lambda) \ge 0$. Moreover, since

$$
\det Cov(u_{-1,.}(t)) = (2\pi)^{p_1} \exp \left\{ \frac{1}{2\pi} \int_{-\pi}^{\pi} \log \det h_{11}(\lambda) d\lambda \right\},
$$

the foregoing arguments imply that the next theorem holds for the overall measure of one-way effect defined in (2.8).

Theorem 2.2 *We have*

$$
M_{v \to u} = \frac{1}{2\pi} \int_{-\pi}^{\pi} M_{v \to u}(\lambda) d\lambda,
\tag{2.17}
$$

where $M_{v \to u} = 0$ if and only if $\bar{u}_{-1,-1}(t)$ belongs to $H\{u(t-1)\}$.

Remark 2.1   Note that the joint process $\{u_{-1,0}(t), v_{0,-1}(t)\}$ of one-way effects has the same information as $\{u(t), v(t)\}$ in the sense that $H\{u_{-1,0}(t), v_{0,-1}(t)\} = H\{u(t), v(t)\}, t \in \mathbb{Z}$.

This is observed as follows. It follows from Lemma 2.4 and the corresponding relation for $u_{-1,0}(t)$ that

$$
\begin{bmatrix} u_{-1,0}(t) \\ v_{0,-1}(t) \end{bmatrix} = \begin{bmatrix} I_{p_1} & -\Sigma_{12}\Sigma_{22}^{-1} \\ -\Sigma_{21}\Sigma_{11}^{-1} & I_{p_2} \end{bmatrix} \begin{bmatrix} u_{-1,-1}(t) \\ v_{-1,-1}(t) \end{bmatrix}
$$

where the matrix on the right-hand side is non-singular because

$$\det \begin{bmatrix} I_{p_1} & -\Sigma_{12}\Sigma_{22}^{-1} \\ -\Sigma_{21}\Sigma_{11}^{-1} & I_{p_2} \end{bmatrix} = \det \Sigma_{22}^{-1} \det\{\Sigma_{22} - \Sigma_{21}\Sigma_{11}^{-1}\Sigma_{12}\}$$

$$= \det \Sigma_{11}^{-1} \det \Sigma_{22}^{-1} \det \Sigma > 0.$$

Consequently, we have

$$H\{u_{-1,0}(t), v_{0,-1}(t)\} = H\{u_{-1,-1}(t), v_{-1,-1}(t)\} = H\{u(t), v(t)\},$$

where the second equality follows from Assumption 2.1, which implies the invertibility of $\Gamma(e^{-i\lambda})\Gamma(0)^{-1}$.

2.4 Alternative Methods for Deriving $M_{v \to u}(\lambda)$

2.4.1 Distributed-Lag Representation Approach

Because $\{u(t)\}$ does not cause $\{v_{0,-1}(t)\}$, we have in view of Definition 2.2 the distributed-lag representation under Assumption 2.1:

$$u(t) = \sum_{j=0}^{\infty} \Pi(j) v_{0,-1}(t-j) + n(t), \quad t \in \mathbb{Z} \qquad (2.18)$$

where the $\Pi(j)$ are $p_1 \times p_2$ matrices and $\{n(t)\}$ is a possibly serially correlated stationary process that is orthogonal to the process $\{v_{0,-1}(t)\}$, whence $h_{11}(\lambda)$, the spectral density of $\{u(t)\}$, is decomposed as

$$h_{11}(\lambda) = h^{(1)}(\lambda) + h^{(2)}(\lambda),$$

where

$$h^{(1)}(\lambda) = \frac{1}{2\pi} \left(\sum_{j=0}^{\infty} \Pi(j) e^{-ij\lambda} \right) \Sigma_{22\cdot1} \left(\sum_{j=0}^{\infty} \Pi(j) e^{-ij\lambda} \right)^{*}.$$

The log-ratio

$$\tilde{M}_{v \to u}(\lambda) = \log\{\det h_{11}(\lambda) / \det h^{(2)}(\lambda)\} \qquad (2.19)$$

and its integration over $[-\pi, \pi]$ are interpretable as measuring, respectively, the frequency-wise and overall projection improvement from the use of the information $H\{v_{0,-1}(t-j), j \in \mathbb{Z}^{+}\}$. In fact, we have the following equivalence:

Theorem 2.3 $\tilde{M}_{v \to u}(\lambda) = M_{v \to u}(\lambda), \; -\pi < \lambda \le \pi.$

Proof It follows then from the Sims representation (2.18) that $n(t)$ is the projection residual of $u(t)$ onto $H\{v_{0,-1}(s); \; s \in \mathbb{Z}\}$ and is representable using $\tilde{h}$ defined in (2.10) as

$$n(t) = \int_{-\pi}^{\pi} e^{it\lambda} \{\Phi_u(d\lambda) - \tilde{h}_{12}(\lambda)\tilde{h}_{22}^{-1}(\lambda)\Phi_{v_{0,-1}}(d\lambda)\}. \tag{2.20}$$

[See Whittle (1984) for such a spectral regression representation as (2.20).] In view of (2.18) and $H\{v_{0,-1}(t-1), u(t-1)\} = H\{v_{0,-1}(t-1), n(t-1)\}$, the residual of $u(t)$ projected onto $H\{v_{0,-1}(t-1), u(t-1)\}$ is equal to the residual of $n(t)$ projected onto $H\{n(t-1)\}$. Therefore, we have

$$\det \tilde{\Sigma}_{11} = (2\pi)^{p_1} \exp\left[\frac{1}{2\pi} \int_{-\pi}^{\pi} \log \det Cov\{\Phi_u(d\lambda) - \tilde{h}_{12}(\lambda)\tilde{h}_{22}^{-1}(\lambda)\Phi_{v_{0,-1}}(d\lambda)\}\right],$$
$$\tag{2.21}$$

where $\tilde{\Sigma}_{11}$ denotes the covariance matrix of the one-step ahead prediction error of $n(t)$ by its own past; whereas for $u(t)$ we have the relation

$$\det \Sigma_{11} = (2\pi)^{p_1} \exp\left[\frac{1}{2\pi} \int_{-\pi}^{\pi} \log \det Cov\{\Phi_u(d\lambda)\}\right]. \tag{2.22}$$

The comparison of (2.21) and (2.22) implies that the prediction improvement by the additional information of $v_{0,-1}(t)$ is given by

$$\tilde{M}_{v \to u} = \log\{\det \Sigma_{11} / \det \tilde{\Sigma}_{11}\}$$

and that the frequency-wise reduction of the variability is given in view of (2.21) and (2.22) by

$$\begin{aligned}
\tilde{M}_{v \to u}(\lambda) &= \log[\det Cov\{\Phi_u(d\lambda)\} / \det Cov\{\Phi_{\Phi_u}(d\lambda) - \tilde{h}_{12}\tilde{h}_{22}^{-1}(\lambda)\Phi_{v_{0,-1}}(d\lambda)\}] \\
&= \log[\det h_{11}(\lambda) / \det\{h_{11}(\lambda) - \tilde{h}_{12}(\lambda)\tilde{h}_{22}^{-1}(\lambda)\tilde{h}_{21}(\lambda)\}] \\
&\equiv \log[\det h_{11}(\lambda) / \det h_{11:2}(\lambda)] \\
&= M_{v \to u}(\lambda). \;\; \square
\end{aligned}$$

2.4.2   Innovation Orthogonalization Approach

The one-way effect measure is constructed in this subsection using the moving average (2.25) below on which the innovation accounting of Sims (1980) is based. The one-way effect measure obtained by the MA representation obtained coincides with

Geweke's measure of linear feedback if the joint process $\{u(t), v(t)\}$ is a stationary autoregressive process. Suppose that the spectral density $h(\lambda)$ has a canonical factorization

$$h(\lambda) = \frac{1}{2\pi} \Gamma(e^{-i\lambda}) \Gamma(e^{-i\lambda})^*,$$

as in (2.6) of the previous section. Choose symmetric positive definite matrices $\Sigma_{11}^{1/2}$ and $\Sigma_{22:1}^{1/2}$ such that $\Sigma_{11}^{1/2} \Sigma_{11}^{1/2} = \Sigma_{11}$ and $\Sigma_{22:1}^{1/2} \Sigma_{22:1}^{1/2} = \Sigma_{22:1} = \Sigma_{22} - \Sigma_{21} \Sigma_{11}^{-1} \Sigma_{12}$. Set

$$\Sigma_{(1)}^{1/2} = \begin{bmatrix} \Sigma_{11}^{1/2} & 0 \\ 0 & \Sigma_{22:1}^{1/2} \end{bmatrix}; \quad \text{hence} \quad \Sigma_{(1)} = \Sigma_{(1)}^{1/2} \Sigma_{(1)}^{1/2} = \begin{bmatrix} \Sigma_{11} & 0 \\ 0 & \Sigma_{22} - \Sigma_{21} \Sigma_{11}^{-1} \Sigma_{12} \end{bmatrix}.$$

Moreover, define the $(p_1 + p_2) \times (p_1 + p_2)$ matrix $A_{(1)}$

$$A_{(1)} \equiv \begin{bmatrix} I_{p_1} & 0 \\ -\Sigma_{21} \Sigma_{11}^{-1} & I_{p_2} \end{bmatrix}$$

and set

$$\Gamma^\dagger(z) = \Gamma(z) \Gamma(0)^{-1} A_{(1)}^{-1} \Sigma_{(1)}^{1/2}. \tag{2.23}$$

Lemma 2.7 $\Gamma^\dagger(z)$ *is a maximal analytic function in the unit disk such that* $\Gamma^\dagger(0) \Gamma^\dagger(0)^* = \Sigma$, *and the spectral density* h *has a canonical factorization*

$$h(\lambda) = \frac{1}{2\pi} \Gamma^\dagger(e^{-i\lambda}) \Gamma^\dagger(e^{-i\lambda})^*.$$

Proof The analyticity of $\Gamma^\dagger(z)$ is evident in view of the construction, and also the maximality follows from the assumption $\Gamma^\dagger(0) \Gamma^\dagger(0)^* = \Sigma$. It follows from (2.23) and $A_{(1)}^{-1} \Sigma_{(1)} A_{(1)}^{-1*} = \Sigma$ that

$$\frac{1}{2\pi} \Gamma^\dagger(e^{-i\lambda}) \Gamma^\dagger(e^{-i\lambda})^* = \frac{1}{2\pi} \Gamma(e^{-i\lambda}) \Gamma(0)^{-1} \Sigma \Gamma(0)^{-1*} \Gamma(e^{-i\lambda})^* = h(\lambda). \qquad \square$$

Define $\bar{\Gamma}(z)$ by $\bar{\Gamma}(z) = \Gamma^\dagger(z) \Sigma_{(1)}^{-1/2}$. Since $\Gamma^\dagger(z)$ is analytic inside the unit circle and so is $\bar{\Gamma}(z)$, the latter has the expansion

$$\bar{\Gamma}(z) = \sum_{j=0}^{\infty} \bar{\Gamma}[j] z^j \tag{2.24}$$

which is convergent on $\{z : |z| < 1\}$ and square integrable *a.e.* on $-\pi < \lambda \leq \pi$ for $z = exp(-i\lambda)$. The matrix $\bar{\Gamma}(e^{-i\lambda})$ is said a frequency response function whereas its time-domain counterpart is the moving average representation by means of the real-matrix coefficient $\bar{\Gamma}[j]$ in (2.24)

$$\begin{bmatrix} u(t) \\ v(t) \end{bmatrix} = \sum_{j=0}^{\infty} \bar{\Gamma}[j] \begin{bmatrix} \varepsilon_1(t-j) \\ \varepsilon_2(t-j) \end{bmatrix} \equiv \sum_{j=0}^{\infty} \begin{bmatrix} \bar{\Gamma}_{11}[j] & \bar{\Gamma}_{12}[j] \\ \bar{\Gamma}_{21}[j] & \bar{\Gamma}_{22}[j] \end{bmatrix} \begin{bmatrix} \varepsilon_1(t-j) \\ \varepsilon_2(t-j) \end{bmatrix} \quad (2.25)$$

where $\{\varepsilon_1(t), \varepsilon_2(t)\}$ is a white noise process with mean 0 and covariance matrix $\Sigma_{(1)}$, and the coefficient $\bar{\Gamma}[j]$ is termed the impulse response function. [See for more detailed expositions on the derivation of moving average representations on the basis of Hilbert space arguments see Rozanov (1967, pp. 28–43) and Hannan (1970, pp. 157–163).] Since

$$\Gamma^{\dagger}(e^{-i\lambda})\Gamma^{\dagger}(e^{-i\lambda})^* = \bar{\Gamma}(e^{-i\lambda})\Sigma_{(1)}\bar{\Gamma}(e^{-i\lambda})^*$$

$$= \begin{pmatrix} \bar{\Gamma}_{11}(e^{-i\lambda}) & \bar{\Gamma}_{12}(e^{-i\lambda}) \\ \bar{\Gamma}_{21}(e^{-i\lambda}) & \bar{\Gamma}_{22}(e^{-i\lambda}) \end{pmatrix} \begin{pmatrix} \Sigma_{11} & 0 \\ 0 & \Sigma_{22} - \Sigma_{21}\Sigma_{11}^{-1}\Sigma_{12} \end{pmatrix} \begin{pmatrix} \bar{\Gamma}_{11}(e^{-i\lambda})^* & \bar{\Gamma}_{21}(e^{-i\lambda})^* \\ \bar{\Gamma}_{12}(e^{-i\lambda})^* & \bar{\Gamma}_{22}(e^{-i\lambda})^* \end{pmatrix},$$

we have

$$h_{11}(\lambda) = \frac{1}{2\pi}\{\bar{\Gamma}_{11}(e^{-i\lambda})\Sigma_{11}\bar{\Gamma}_{11}(e^{-i\lambda})^* + \bar{\Gamma}_{12}(e^{-i\lambda})\Sigma_{22:1}\bar{\Gamma}_{12}(e^{-i\lambda})^*\}.$$

Now, define the measure $\bar{M}_{v \to u}(\lambda)$ by

$$\bar{M}_{v \to u}(\lambda) = \log\left[\det h_{11}(\lambda) / \det\{\frac{1}{2\pi}\bar{\Gamma}_{11}(e^{-i\lambda})\Sigma_{11}\bar{\Gamma}_{11}(e^{-i\lambda})^*\}\right]. \quad (2.26)$$

Notice that the definition does not depend on the choice of $\Sigma_{(1)}^{1/2}$ and $\Gamma(e^{-i\lambda})$. The measure $\bar{M}_{v \to u}(\lambda)$ thus defined in (2.26) turns out to be equivalent to $M_{v \to u}(\lambda)$ of (2.16).

Theorem 2.4 $\bar{M}_{v \to u}(\lambda) = M_{v \to u}(\lambda).$

Proof It follows from (2.25) and the maximality of $\bar{\Gamma}(z)$ that the residuals $u_{-1,-1}(t)$ and $v_{-1,-1}(t)$ have a representation

$$\begin{bmatrix} u_{-1,-1}(t) \\ v_{-1,-1}(t) \end{bmatrix} = \begin{bmatrix} \bar{\Gamma}_{11}[0] & \bar{\Gamma}_{12}[0] \\ \bar{\Gamma}_{21}[0] & \bar{\Gamma}_{22}[0] \end{bmatrix} \begin{bmatrix} \varepsilon_1(t) \\ \varepsilon_2(t) \end{bmatrix}.$$

Hence,

$$\begin{aligned} v_{0,-1}(t) &= v_{-1,-1}(t) + A_{21}u_{-1,-1}(t) \\ &= \{\bar{\Gamma}_{21}[0] + A_{21}\bar{\Gamma}_{11}[0]\}\varepsilon_1(t) + \{\bar{\Gamma}_{22}[0] + A_{21}\bar{\Gamma}_{12}[0]\}\varepsilon_2(t). \quad (2.27) \end{aligned}$$

Because $\bar{\Gamma}[0] = A_{(1)}^{-1}$ in view of (2.23), we have

$$\begin{bmatrix} I_{p_1} & 0 \\ A_{21} & I_{p_2} \end{bmatrix} \begin{bmatrix} \bar{\Gamma}_{11}[0] & \bar{\Gamma}_{12}[0] \\ \bar{\Gamma}_{21}[0] & \bar{\Gamma}_{22}[0] \end{bmatrix} = I_p \tag{2.28}$$

It follows from (2.27) and (2.28) that $v_{0,-1}(t) = \varepsilon_2(t)$. Therefore, the residual of the projection of $u(t)$ onto $H\{v_{0,-1}(\infty)\} = H\{\varepsilon_2(\infty)\}$ is equal to $\sum_{j=0}^{\infty} \bar{\Gamma}_{11}[j]\varepsilon_1(t-j)$ in view of (2.25). It follows from the definition of $M_{v \to u}(\lambda)$ in (2.16) that

$$M_{v \to u}(\lambda) = \log \frac{\det h_{11}(\lambda)}{\det\{\frac{1}{2\pi} \sum_{j=0}^{\infty} \bar{\Gamma}_{11}[j]e^{-ij\lambda}\}\Sigma_{11}\{\frac{1}{2\pi} \sum_{j=0}^{\infty} \bar{\Gamma}_{11}[j]e^{-ij\lambda}\}^*} = \bar{M}_{v \to u}(\lambda).$$

This equality holds because $h_{11:2}(\lambda) = h_{11}(\lambda) - 2\pi h_{12}(\lambda)\Sigma_{22:1}^{-1}h_{21}(\lambda)$ in the bracket on the right-hand side of (2.16) is a spectral density of the residual process of the projection $u(t)$ onto $H\{v_{0,-1}(\infty)\}$. □

Suppose that the process $\{u(t), v(t)\}$ has the stationary autoregressive representation

$$\begin{bmatrix} u(t) \\ v(t) \end{bmatrix} = \sum_{j=1}^{\infty} H[j] \begin{bmatrix} u(t-j) \\ v(t-j) \end{bmatrix} + \begin{bmatrix} \varepsilon_1(t) \\ \varepsilon_2(t) \end{bmatrix}$$

where $\{\varepsilon_1(t), \varepsilon_2(t)\}$ is a white noise process with covariance matrix Σ. Set $H(z) = I_{p_1+p_2} - \sum_{j=1}^{\infty} H[j]z^j$, and define $\Gamma_{(1)}(z) \equiv H(z)^{-1}A_{(1)}^{-1}\Sigma_{(1)}^{1/2}$. Then, $\Gamma_{(1)}(z)$ is nothing but a version of the canonical factor $\Gamma(e^{-i\lambda})$ in (2.6) and, thus, possesses the properties prescribed in Lemma 2.2. Geweke's measure of linear feedback is then constructed by (2.26) on the basis of this version. Chapter 3 expounds on the construction of the measure for the vector ARMA model.

2.5 Measures of Association and Reciprocity

This subsection introduces measures of association and reciprocity as additional measures to characterize the interdependence structure of the processes $\{u(t)\}$ and $\{v(t)\}$. This subsection also shows that the measure of association is decomposed into the measures of one-way effect and reciprocity.

Recall that $\bar{u}'_{,\infty}(t)$ and $u'_{,\infty}(t)$ are the projections of $u(t)$ onto $H\{v_{0,-1}(\infty)\}$ and its residual and that $\bar{v}'_{\infty,.}(t)$ and $v'_{\infty,.}(t)$ are defined similarly; hence, we have

$$u(t) = u'_{,\infty}(t) + \bar{u}'_{,\infty}(t); \quad v(t) = v'_{\infty,.}(t) + \bar{v}'_{\infty,.}(t). \tag{2.29}$$

Lemma 2.8 *For the decomposition (2.29), we have*

$$u'_{.,\infty}(t) \in H\{u_{-1,-1}(t)\}; \quad \bar{u}'_{.,\infty} \in H\{v_{0,-1}(t-1)\},$$

and

$$v'_{\infty,.}(t) \in H\{v_{-1,-1}(t)\}; \quad \bar{v}'_{\infty,.}(t) \in H\{u_{-1,0}(t-1)\}.$$

Proof The proof for $\{u(t)\}$ is shown by the following steps:

a. We have the equalities $H\{u(t), v(t)\} = H\{u_{-1,-1}(t), v_{-1,-1}(t)\} = H\{u_{-1,-1}(t)\} \oplus H\{v_{0,-1}(t)\}$ and, hence, $u(t) \in H\{u_{-1,-1}(t)\} \oplus H\{v_{0,-1}(t)\}$. However, because $u'_{.,\infty}(t)$ is the projection residual of $u(t)$ onto $H\{v_{0,-1}(\infty)\}$, $u'_{.,\infty}(t) \perp H\{v_{0,-1}(t)\}$. Therefore, we have $u'_{.,\infty}(t) \in H\{u_{-1,-1}(t)\}$.
b. Because $\bar{u}'_{.,\infty}(t)$ is the projection of $u(t)$ onto $H\{v_{0,-1}(\infty)\}$; however, because $u(t-j) \perp v_{0,-1}(t), \ j \geq 0$, we have $\bar{u}'_{.,\infty}(t) \in H\{v_{0,-1}(t-1)\}$.

A similar train of arguments holds for $\{v(t)\}$.   □

Denote the joint spectral density of the process $\{u'_{.,\infty}(t), v'_{\infty,.}(t)\}$ by

$$h'(\lambda) = \begin{bmatrix} h'_{11}(\lambda) & h'_{12}(\lambda) \\ h'_{21}(\lambda) & h'_{22}(\lambda) \end{bmatrix}.$$

The next lemma is straightforward in view of the construction, and the proof is omitted. [See Lemma 2.5. for allied notations.]

Lemma 2.9 *The spectral densities $h'_{11}(\lambda), h'_{22}(\lambda)$ of the processes $\{u'_{.,\infty}(t)\}$ and $\{v'_{\infty,.}(t)\}$, respectively, are given as follows:*

$$h'_{11}(\lambda) \equiv h_{11}(\lambda) - 2\pi \tilde{h}_{12}(\lambda) \Sigma_{22:1}^{-1} \tilde{h}_{21}(\lambda);$$
$$h'_{22}(\lambda) \equiv h_{22}(\lambda) - 2\pi \tilde{h}_{21}(\lambda) \Sigma_{11:2}^{-1} \tilde{h}_{12}(\lambda).$$

Define the measure of association at frequency λ between the two processes $\{u(t)\}$ and $\{v(t)\}$ by

$$M_{u,v}(\lambda) \equiv \log[\det h_{11}(\lambda) \det h_{22}(\lambda) / \det h'(\lambda)] \qquad (2.30)$$

and define the measure of reciprocity at frequency λ by

$$M_{u.v}(\lambda) \equiv \log[\det h'_{11}(\lambda) \det h'_{22}(\lambda) / \det h'(\lambda)]. \qquad (2.31)$$

It follows from the nonnegative definiteness of $h_{11}(\lambda) - h'_{11}(\lambda)$ and $h_{22}(\lambda) - h'_{22}(\lambda)$ that $M_{u,v}(\lambda) \geq 0$. It is evident that $M_{u,v}(\lambda) \geq 0$. Define the corresponding overall measure $M_{u,v}(\lambda)$ and $M_{u,v}(\lambda)$ by the integration of the respective frequency-wise measure,

$$M_{u,v} \equiv \frac{1}{2\pi} \int_{-\pi}^{\pi} M_{u,v}(\lambda)d\lambda; \quad M_{u,v} \equiv \frac{1}{2\pi} \int_{-\pi}^{\pi} M_{u,v}(\lambda)d\lambda.$$

The following equalities hold between those measures and the measures of the one-way effect.

Theorem 2.5 *We have the equality*

$$M_{u,v}(\lambda) = M_{u \to v}(\lambda) + M_{u,v}(\lambda) + M_{v \to u}(\lambda) \tag{2.32}$$

and, consequently,

$$M_{u,v} = M_{u \to v} + M_{u,v} + M_{v \to u}. \tag{2.33}$$

Proof In view of $M_{v \to u}(\lambda) = \log\{\det h_{11}(\lambda)/\det h'_{11}(\lambda)\}$, the relation (2.32) follows from the equality

$$\log\{\det h_{11}(\lambda)\det h_{22}(\lambda)/\det h'(\lambda)\}$$
$$= \log\{\det h_{11}(\lambda)/\det h'_{11}(\lambda)\} + \log\{\det h'_{11}(\lambda)\det h'_{22}(\lambda)/\det h'(\lambda)\}$$
$$+ \log\{\det h_{22}(\lambda)/\det h'_{22}(\lambda)\}.$$

The relation (2.33) is then evident. □

To characterize the interdependency between second-order stationary processes $\{u(t)\}$ and $\{v(t)\}$, Gel'fand and Yaglom (1956) introduced the measure of information $\tilde{M}_{u,v}$ defined by

$$\tilde{M}_{u,v} = \log[\det\{Cov(u_{-1,.}(t))\}\det\{Cov(v_{.,-1}(t))\}/\det\{Cov(u_{-1,-1}(t), v_{-1,-1}(t))\}]$$
$$= \frac{1}{2\pi}\int_{-\pi}^{\pi} \log[\det h_{11}(\lambda)\det h_{22}(\lambda)/\det h(\lambda)]d\lambda. \tag{2.34}$$

Geweke (1982) termed the quantity $\tilde{M}_{u,v} \equiv \log\{\det \Sigma_{11} \cdot \det \Sigma_{22}/\det \Sigma\}$ the measure of instantaneous feedback; see also Theorem 3.3 of Sect. 3.3.1. Note that $M_{u,v} = \tilde{M}_{u,v}$ if $\{u(t), v(t)\}$ is a white noise process. The measures of association and reciprocity coincide with those measures under certain conditions.

Theorem 2.6 (i) *If* $H\{u'_{.,\infty}(t), v'_{\infty,.}(t)\} = H\{u(t), v(t)\}, t \in \mathbb{Z}$, *we have* $M_{u,v} = \tilde{M}_{u,v}$. (ii) *If* $H\{u'_{.,\infty}(t)\} = H\{u_{-1,-1}(t)\}$ *and* $H\{v'_{\infty,.}(t)\} = H\{v_{-1,-1}(t)\}$, *we have* $M_{u,v} = \tilde{M}_{u,v}$ *and* $M_{u,v} = \tilde{M}_{u,v}$.

Proof Suppose that $H\{u'_{.,\infty}(t), v'_{\infty,.}(t)\} = H\{u(t), v(t)\}$. Then, the residual of the projection of $u'_{.,\infty}(t)$ onto $H\{u'_{.,\infty}(t-1), v'_{\infty,.}(t-1)\}$ is $u_{-1,-1}(t)$ because $u(t) = u'_{.,\infty}(t) + \bar{u}'_{.,\infty}(t)$ and $\bar{u}'_{.,\infty}(t) \in H\{v_{0,-1}(t-1)\} \subset H\{u(t-1), v(t-1)\} \subset H\{u'_{.,\infty}(t-1), v'_{\infty,.}(t-1)\}$. Similarly, the residual of the projection of $v'_{\infty,.}(t)$ onto $H\{u'_{.,\infty}(t-1), v'_{\infty,.}(t-1)\}$ is $v_{-1,-1}(t)$. Hence, the one-step ahead prediction errors of the process $\{u(t), v(t)\}$ and of the process $\{u'_{.,\infty}, v'_{\infty,.}(t)\}$ are the same, whence

$$\frac{1}{2\pi} \int_{-\pi}^{\pi} \log \det h'(\lambda) d\lambda = \frac{1}{2\pi} \int_{-\pi}^{\pi} \log \det h(\lambda) d\lambda. \tag{2.35}$$

Consequently, the equality $M_{u,v} = \tilde{M}_{u,v}$ follows. Regarding proposition (ii), note that the assumptions imply that $H\{u'_{\cdot,\infty}(t), v'_{\infty,\cdot}(t)\} = H\{u(t), v(t)\}$, whence (2.35) and $M_{u,v} = \tilde{M}_{u,v}$ hold. Moreover, because then $H\{u'_{\cdot,\infty}(t), v_{0,-1}(t)\} = H\{u_{-1,-1}(t), v_{-1,-1}(t)\}$, it follows that $u'_{-1,-1}(t) = u_{-1,-1}(t)$ and $Cov\{u'_{-1,-1}(t)\} = \Sigma_{11}$. In the same way, we have $Cov\{v'_{-1,-1}(t)\} = \Sigma_{22}$. Hence, the equality $M_{u,v} = \tilde{M}_{u,v}$ follows from (2.30) and (2.31). $\qquad\square$

See Theorem 3.3 for another proof of $M_{u,v} = \tilde{M}_{u,v}$ in (ii) of Theorem 2.6 under a general condition.

Set $A = (-\Sigma_{21}\Sigma_{11}^{-1}, I_{p_2}); \quad B = (I_{p_1}, -\Sigma_{12}\Sigma_{22}^{-1})$, where

$$\Sigma = \begin{bmatrix} \Sigma_{11} & \Sigma_{12} \\ \Sigma_{21} & \Sigma_{22} \end{bmatrix}$$

is the covariance of the one-step ahead prediction error of the process $\{u(t), v(t)\}$. Also set $\Sigma_{11:2} \equiv \Sigma_{11} - \Sigma_{12}\Sigma_{22}^{-1}\Sigma_{21}$ and $\Sigma_{22:1} \equiv \Sigma_{22} - \Sigma_{21}\Sigma_{11}^{-1}\Sigma_{12}$. As is seen in the preceding arguments, evaluation of the frequency-wise measures of interdependence requires an explicit representation of $h'(\lambda)$.

Theorem 2.7 *The spectral density $h'(\lambda)$ is represented as follows:*

$$h'_{11}(\lambda) = h_{11}(\lambda) - 2\pi h_{1\cdot}(\lambda)\Gamma(e^{-i\lambda})^{-1*}\Gamma(0)^*A^*\Sigma_{22:1}^{-1}A\Gamma(0)\Gamma(e^{-i\lambda})^{-1}h_{\cdot1}(\lambda),$$

$$h'_{22}(\lambda) = h_{22}(\lambda) - 2\pi h_{2\cdot}(\lambda)\Gamma(e^{-i\lambda})^{-1*}\Gamma(0)^*B^*\Sigma_{11:2}^{-1}B\Gamma(0)\Gamma(e^{-i\lambda})^{-1}h_{\cdot2}(\lambda),$$

$$h'_{12}(\lambda) = h_{12}(\lambda) - 2\pi h_{1\cdot}(\lambda)\Gamma(e^{-i\lambda})^{-1*}\Gamma(0)^*(A^*\Sigma_{22:1}^{-1}A + B^*\Sigma_{11:2}^{-1}B)\Gamma(0)\Gamma(e^{-i\lambda})^{-1}h_{\cdot2}(\lambda)$$
$$+ 2\pi\tilde{h}_{12}(\lambda)\Sigma_{22:1}^{-1}(-\Sigma_{21} + \Sigma_{21}\Sigma_{11}^{-1}\Sigma_{12}\Sigma_{22}^{-1}\Sigma_{21})\Sigma_{11:2}^{-1}\check{h}_{12}(\lambda). \tag{2.36}$$

Proof The expression (2.36) is the corrected version of Theorem 4.3 of Hosoya (1991), and the proof is omitted. Theorem 3.3 of Chap. 3 presents another representation and the derivation. $\qquad\square$

Remark 2.2 In general, the measure of association $M_{u,v}$ is not equal to the Gel'fand–Yaglom measure $\tilde{M}_{u,v}$ given in (2.34). Example 2.2 of Sect. 2.6 provides a case in which $\tilde{M}_{u,v} > M_{u,v}$.

Remark 2.3 The notion of Akaike (1968) of relative power contribution (RPC) can be employed in an extended form for the purpose of representation of $M_{v\to u}(\lambda)$. Let $\bar{h}'_{11}(\lambda)$ be the spectral density of the process $\{\bar{u}'_{\cdot,\infty}(t)\}$; hence, $\bar{h}'_{11}(\lambda) = h_{11}(\lambda) - h'_{11}(\lambda)$, and let $r_1(\lambda) \geq \cdots \geq r_{p_1}(\lambda)$ be the eigenvalues of the matrix $\bar{h}'_{11}(\lambda)h_{11}(\lambda)^{-1}$. Evidently, all of the $r_j(\lambda)$ are nonnegative real values and $r_1(\lambda) \leq 1$. Define $RPC_{v\to u}(\lambda)$, the relative power contribution of $\{v(t)\}$ to $\{u(t)\}$, by the diagonal $p_1 \times p_1$ matrix with $r_j(\lambda)$ for the (j, j) element. Because

$$M_{v \to u}(\lambda) = \log\{\det h_{11}(\lambda)/\det h'_{11}(\lambda)\},$$

the one-way frequency-wise measure is related to Akaike's RPC by

$$M_{v \to u}(\lambda) = -\log \det\{I_{p_1} - RPC_{v \to u}(\lambda)\}.$$

See Tanokura and Kitagawa (2015) for an allied study.

Remark 2.4 The overall measures of association, one-way effect, and reciprocity are defined on the basis of only the one-step ahead prediction error of the processes $\{u(t), v(t)\}$, $\{u'_{.,\infty}(t), v'_{\infty,.}(t)\}$ or of their component processes. Also, the processes $\{u_{-1,0}(t)\}$ and $\{v_{0,-1}(t)\}$ are defined in terms of the projection errors. Therefore, as long as we work with these prediction error-related concepts, the assumption of stationarity of the original process can be dispensed with; see also Theorem 2.1 of Sect. 2.2. In particular, the decomposition $M_{u,v} = M_{u \to v} + M_{u \cdot v} + M_{v \to u}$ is still valid for non-stationary second-order processes. Also valid is the proposition that $M_{v \to u}$ is the log-ratio of $\det Cov(u_{-1,.}(t))$ to $\det Cov(u'_{-1,-1}(t))$, which equals the log-ratio of the determinants of the one-step ahead prediction errors of $\{u(t)\}$ and $\{u'_{.,\infty}(t)\}$.

2.6 Examples

This section shows two cases of the bivariate moving average process. The first case is the non-invertible process for which $M_{v \to u}(\lambda)$ is evaluated. The second case is the situation in which $F_{v \to u} > M_{v \to u}$; hence, the two measures diverge.

Example 2.1 Consider a bivariate process $\{u(t), v(t)\}$, $t \in \mathbb{Z}$, which is generated by

$$u(t) = \varepsilon(t) - \varepsilon(t - 1) + a\eta(t - 1) + b\eta(t - 2)$$
$$v(t) = \eta(t) + c\eta(t - 1)$$

where we assume that $|c| \le 1$ and $\{\varepsilon(t), \eta(t)\}$ is a white noise process such that $E(\varepsilon(t)) = E(\eta(t)) = 0$, $Var(\varepsilon(t)) = Var(\eta(t)) = 1$, and $Cov(\varepsilon(t), \eta(t)) = 0$. Setting

$$\Gamma(z) = \begin{pmatrix} 1 - z & az + bz^2 \\ 0 & 1 + cz \end{pmatrix},$$

we have the spectral density matrix of the $\{u(t), v(t)\}$, which is given by

$$h(\lambda) = \frac{1}{2\pi}\Gamma(e^{-i\lambda})\Gamma(e^{-i\lambda})^*$$

$$= \frac{1}{2\pi}\left[\begin{array}{cc} |1 - e^{-i\lambda}|^2 + |a + be^{-i\lambda}|^2 & (ae^{-i\lambda} + be^{-2i\lambda})(1 + ce^{-i\lambda})^* \\ (ae^{-i\lambda} + be^{-2i\lambda})^*(1 + ce^{-i\lambda}) & |1 + ce^{-i\lambda}|^2 \end{array}\right]$$

where $\Gamma(z)$ is a canonical factor. Thus, we have the equality

$$(2\pi)^2 \exp\left\{\frac{1}{2\pi}\int_{-\pi}^{\pi} \log \det h(\lambda)d\lambda\right\} = |\Gamma(0)|^2.$$

Because $\Gamma(z)$ is maximal, $\{\varepsilon(t), \eta(t)\}$ constitutes an innovation process for $\{u(t), v(t)\}$; namely, we have $\varepsilon(t) = u_{-1,-1}(t)$ and $\eta(t) = v_{-1,-1}(t)$. Then, because $v_{0,-1}(t) = \eta(t)$ in view of Lemma 2.4, by applying the distributed-lag approach of Sect. 2.4.1, the spectral density of the residual process $\{n(t)\}$ is given by $h^{(2)}(\lambda) = |1 - e^{-i\lambda}|^2/(2\pi)$. Hence, it follows from (2.19) that

$$M_{v\to u}(\lambda) = \log \frac{|1 - e^{-i\lambda}|^2 + |a + be^{-i\lambda}|^2}{|1 - e^{-i\lambda}|^2}.$$

Therefore, $\{v(t)\}$ does not cause $\{u(t)\}$ if and only if $a = b = 0$.

Example 2.2   Suppose that a bivariate process $\{u(t), v(t)\}, t \in \mathbb{Z}$, is generated by

$$u(t) = \varepsilon(t) + a\varepsilon(t - 1) + b\eta(t - 1)$$
$$v(t) = \eta(t) + c\varepsilon(t - 1) + d\eta(t - 1)$$

where the white noise process $\{\varepsilon(t), \eta(t)\}$ is specified as in Example 2.1. Set

$$\Gamma(z) = \begin{pmatrix} 1 + az & bz \\ cz & 1 + dz \end{pmatrix}.$$

Suppose that the parameters satisfy that $ad = -1$ and $bc = -3/2$, then the zeros of $\det \Gamma(z)$ are the roots of the quadratic equation

$$\det \Gamma(z) = z^2/2 + (a + d)z + 1 = 0.$$

Suppose, for example, that $b = 0.1$, and $0.7 \le a \le 1.6$, then all the zeros z_0 of $\det \Gamma(z)$ satisfy $1.04 \le |z_0| \le 1.47$; hence, they are outside of the unit circle. If all zeros of $\det \Lambda(z)$ are outside the unit circle, the spectral density $h(\lambda)$ of the process $\{u(t), v(t)\}$ is given in terms of the canonical factor $\Gamma(e^{-i\lambda})$ by

$$h(\lambda) = \frac{1}{2\pi}\Gamma(e^{-i\lambda})\Gamma(e^{-i\lambda})^*.$$

For such a canonical factor $\Gamma(z)$, we have $v_{0,-1}(t) = \eta(t)$, and thus $M_{v\to u}(\lambda)$ is expressed as

$$M_{v \to u}(\lambda) = \log \frac{|1 + ae^{-i\lambda}|^2 + b^2}{|1 + ae^{-i\lambda}|^2},$$

whereas the Geweke measure is given by

$$F_{v \to u} = \frac{1}{2\pi} \int_{-\pi}^{\pi} \log\{|1 + ae^{-i\lambda}|^2 + b^2\} d\lambda - \log Var(\varepsilon(t))$$

because $H\{u(t), v(t)\} = H\{\varepsilon(t), \eta(t)\}$. Moreover, because $\log Var(\varepsilon(t)) = 0$ and

$$\int_0^{\pi} \log\{|1 + 2a\cos(\lambda) + a^2|\} d\lambda = 2\pi \log |a|,$$

if $a = 1.5$ for example, the difference between the two measures is equal to

$$F_{v \to u} - M_{v \to u} = \frac{1}{2\pi} \int_{-\pi}^{\pi} \log\{|1 + a^{-i\lambda}|^2\} d\lambda = 2 \log |a| = 2 \log 1.5 > 0.$$

References

Akaike, H. (1968). On the use of a linear model for the identification of feedback system. *Annals of the Institute of Statistical Mathematics, 20*, 425–439.

Florens, J. P., & Mouchart, M. (1982). A note on noncausality. *Econometrica, 50*, 583–592.

Gel'fand, I. M., & Yaglom, A. M. (1959). Calculation of the amount of information about a random function contained in another such function. *American Mathematical Society Translation Series, 2*(12), 199–246.

Geweke, J. (1982). Measurement of linear dependence and feedback between multiple time series. *Journal of the American Statistical Association, 77*, 304–324.

Granger, C. W. J. (1963). Economic process involving feedback. *Information and Control, 6*, 28–48.

Granger, C. W. J. (1969). Investigating causal relations by econometric methods and cross-spectral methods. *Econometrica, 37*, 424–438.

Hannan, E. J. (1970). *Multiple Time Series*. New York: Wiley.

Hosoya, Y. (1977). On Granger condition for non-causality. *Econometrica, 45*, 1735–1736.

Hosoya, Y. (1991). The decomposition and measurement of the interdependency between second-order stationary processes. *Probability Theory and Related Fields, 88*, 429–444.

Pierce, D. A. (1979). R^2 measures for time series. *Journal of American Statistical Association, 74*, 901–910.

Rozanov, Y. A. (1967). *Stationary Random Processes*. San Francisco: Holden Day.

Sims, C. A. (1972). Money, income and causality. *American Economic Review, 62*, 540–552.

Tanokura, Y., & Kitagawa, G. (2015). *Indexation and Causation of Financial Markets: Nonstationary Time Series Method*. Tokyo: Springer.

Whittle, P. (1984). *Prediction and Regulation by Linear Least-Square Method* (2nd ed.). Oxford: Basil Blackwell Publisher.

Chapter 3
Representation of the Partial Measures

Abstract This chapter extends the measures introduced in the previous chapter to partial measures in the presence of third-series involvement. Third-series intervention is known to sometimes incur phenomena such as spurious or indirect causality attributable to possible feedback from the series. To address the problem, this chapter introduces an operational way to define the partial causality and allied concepts between a pair of processes. The third-effect elimination is of the one-way effect component of the third series from a pair of subject-matter series to preserve the inherent feedback structure of the pair of interest.

Keywords Canonical factorization · Cointegrated process · Partial measures of interdependence · Simple measures of interdependence · Spurious causality · Third-series presence · Unit-root process · Vector ARMA process

3.1 Introduction

The chapter is organized as follows. Section 3.2 illustrates using examples how the proposed elimination method works and provides an extension of Sims' characterization of Granger non-causality to the case of a third-series presence (Theorem 3.1). The section also shows that the proposed partial concept avoids Hsiao's spurious causality (Theorem 3.2). Section 3.3 expounds on the background for the concept of the one-way effect via the Sims representation of Granger's non-causality and discusses how to construct the partial causal measures between a pair of non-deterministic stationary processes on the basis of the elimination from this pair of the one-way effect by a third process. Dealing with a class of possibly non-stationary reproducible processes, Sect. 3.4 extends the partial measures to that class and explains how to construct those measures in non-stationary cointegration processes.

© The Author(s) 2017   45
Y. Hosoya et al., *Characterizing Interdependencies of Multiple Time Series*,
JSS Research Series in Statistics, DOI 10.1007/978-981-10-6436-4_3

3.2   Third-Series Involvement

Suppose that we have a system of zero-mean second-order three vector processes $\{x(t), y(t), z(t); t \in \mathbb{Z}\}$. The notations of the previous chapter are extended to the three-variable system. Let H be the Hilbert space, which is the closure in the mean square of the linear hull of $\{x_j(t); \ t \in \mathbb{Z}, \ j = 1, \cdots, p_1\}$, $\{y_k(t); \ t \in \mathbb{Z}, \ k = 1, \cdots, p_2\}$, and $\{z_l(t); \ t \in \mathbb{Z}, \ l = 1, \cdots, p_3\}$ in the space of all random variables with a finite variance, where $x_j(t)$ denotes the jth component of the vector $x(t)$. For brevity, $H\{x(t_1-j), y(t_2-j), z(t_3-j); \ j \in \mathbb{Z}^{0+}\}$ is written as $H\{x(t_1), y(t_2), z(t_3)\}$ and $H\{x(j); \ j \in \mathbb{Z}\}$ is written as $H\{x(\infty)\}$. As in the previous chapter, the projection of a random vector $w = \{w_j; \ j = 1, \cdots, s\}$ onto $H(\cdot)$, a closed subspace of H, implies a component-wise orthogonal projection. Namely, if $\bar{w}_j$ is the projection of w_j onto $H(\cdot)$, then the projection of w onto $H(\cdot)$ implies the vector $\bar{w}$, whose jth component is $\bar{w}_j$, and the vector $w - \bar{w}$ is called the projection residual.

If the focus is on the relationship of a particular pair $\{x(t), y(t)\}$ in the system $\{x(t), y(t), z(t)\}$, there is a standard method to eliminate the effect attributable to $\{z(t)\}$. Namely, the method bases the analysis on partial serial covariances or a partial spectrum defined for the derived series $\{x_{\cdot,\cdot,\infty}(t)\}$ and $\{y_{\cdot,\cdot,\infty}(t)\}$, where $x_{\cdot,\cdot,\infty}(t)$ and $y_{\cdot,\cdot,\infty}(t)$ denote, respectively, the projection residuals of $x(t)$ and $y(t)$ onto $H\{z(\infty)\}$, the closed linear space generated by $\{z(t), t \in \mathbb{Z}\}$. To address the causal relationships between two processes with the z-effect eliminated, Granger (1969), for example, proposed the use of a partial cross-spectrum using the residual processes $\{x_{\cdot,\cdot,\infty}(t)\}$ and $\{y_{\cdot,\cdot,\infty}(t)\}$. The term "partial" is used as a straightforward extension of the use of multivariate analysis. Unfortunately, that usage does not extend to a causal analysis of a time series, for which temporal order is crucial. The total elimination of the z-effect by projection onto $H\{z(\infty)\}$ accompanies the distortion of the original feedback relations between $\{x(t)\}$ and $\{y(t)\}$, as subsequent examples show.

For the process $\{x(t), y(t), z(t)\}$, the one-way effect component of $z(t)$, which is denoted by $z_{0,0,-1}(t)$, is defined as the projection residual of $z(t)$ onto $H\{x(t), y(t), z(t-1)\} \equiv H\{x(t+1-j), y(t+1-j), z(t-j); \ j \in \mathbb{Z}^+\}$. Denoting by $u(t)$ and $v(t)$ the projection residuals of $x(t)$ and $y(t)$ onto $H\{z_{0,0,-1}(\infty)\}$, respectively, we identify the partial interdependencies between $\{x(t)\}$ and $\{y(t)\}$ in the presence of $\{z(t)\}$, with the corresponding relations between $\{u(t)\}$ and $\{v(t)\}$; see Hosoya (2001). To distinguish an interdependency concept that focuses only on a pair of processes from the partial version that accounts for a third series, the former concept is said to be simple in the sequel; namely, the simple causality, in this book, implies one that does not take a third series into account.

Example 3.1   Let $\{\varepsilon(t)\}$ and $\{\eta(t)\}$ be scalar-valued white noise processes that are orthogonal to each other, and let $\{x(t), y(t), z(t)\}$ be generated by

$$x(t) = \varepsilon(t), \quad y(t) = x(t-1) + \eta(t) \quad \text{and} \quad z(t) = x(t-2).$$

In this case, the one-way Granger causality is directed from $\{x(t)\}$ to $\{y(t)\}$ and $\{z(t)\}$. However, the projection onto $H\{z(\infty)\}$ produces the residuals $x_{\cdot,\cdot,\infty}(t) = 0$

and $y_{.,.,\infty}(t) = \eta(t)$, respectively, such that x does not partially cause y if the total z-effect is eliminated. Therefore, the elimination of the total effect of $\{z(t)\}$ produces an improper picture of interdependency.

Example 3.2 Suppose that a trivariate process $\{x(t), y(t), z(t)\}$ is generated by

$$\begin{cases} x(t) = \alpha y(t-1) + \varepsilon(t) \\ y(t) = \eta(t) \\ z(t) = \beta y(t-1) + \xi(t) + \gamma \xi(t-1) \end{cases}$$

where α, β, γ are nonzero and $|\gamma| < 1$; $\{\varepsilon(t)\}$, $\{\eta(t)\}$ and $\{\xi(t)\}$ are mutually orthogonal scalar-valued white noise processes. Although, in this system, $\{y(t)\}$ appears to cause one-sidedly $\{x(t)\}$ in the Granger sense, if the total z-effect is eliminated as in Example 3.1, $\{x_{.,.,\infty}(t)\}$ turns out to cause $\{y_{.,.,\infty}(t)\}$ because both $x_{.,.,\infty}(t)$ and $y_{.,.,\infty}(t)$ contain the $\xi(t)$ and $\xi(t-1)$ elements. In contrast, in terms of the one-way effect elimination, because $z_{0,0,-1}(t) = \xi(t)$, $u(t) = \alpha v(t-1) + \varepsilon(t)$ and $v(t) = \eta(t)$, $v(t)$ remains to cause $u(t)$ one-sidedly, where $u(t)$ and $v(t)$ denote the $z_{0,0,-1}$-effect eliminated $x(t)$ and $y(t)$, respectively.

Geweke (1984) proposed his conditional measures of linear feedback between $\{x(t)\}$ and $\{y(t)\}$ by conditioning $x(t)$ and $y(t)$ on $H\{z(t-1)\}$.

Example 3.3 Let $\{\varepsilon(t)\}$, $\{\eta(t)\}$, and $\{\xi(t)\}$ be mutually orthogonal white noise processes such that the variance ratio $\sigma_\xi^2/\sigma_\eta^2$ is very small. Suppose that the three series $\{x(t)\}$, $\{y(t)\}$, and $\{z(t)\}$ are given by

$$x(t) = y(t-2) + \varepsilon(t), \quad y(t) = \eta(t) \quad \text{and} \quad z(t) = y(t-1) + \xi(t)$$

such that $\{y(t)\}$ seems to cause one-sidedly $\{x(t)\}$ in the Granger sense. Geweke's elimination of the z-effect produces

$$x_{.,.,-1}(t) = y(t-2)\{1 - \frac{\sigma_\eta^2}{\sigma_\eta^2 + \sigma_\xi^2}\} + \varepsilon(t) - \frac{\sigma_\eta^2}{\sigma_\eta^2 + \sigma_\xi^2}\xi(t-1), \quad y_{.,.,-1}(t) = y(t),$$

where the effect attributed to $y(t)$ in $x(t)$ is greatly diminished. In contrast, the projection onto $H\{z_{0,0,-1}(\infty)\} = H\{\xi(\infty)\}$ does not produce this type of distortion.

The next example was given by Hsiao (1982) as a typical case of spurious causality.

Example 3.4 Let $\{\varepsilon(t), \eta(t), \xi(t)\}$ be specified in the same manner as in Example 3.3 and suppose that $\{x(t), y(t), z(t)\}$ is generated by

$$x(t) = \varepsilon(t) + 0.5\eta(t-1), \quad y(t) = \xi(t) \quad, \text{and} \quad z(t) = \eta(t) + 0.5\xi(t-1).$$

This is an invertible MA process, having the AR representation:

$$\begin{cases} x(t) = -0.25y(t-2) + 0.5z(t-1) + \varepsilon(t) \\ y(t) = \xi(t) \\ z(t) = 0.5y(t-1) + \eta(t). \end{cases}$$

Because the process $\{y(t)\}$ is orthogonal with $\{x(t)\}$ as a simple relation, $\{y(t)\}$ does not seem to cause $\{x(t)\}$. However, if the entire past of the third process $\{z(t)\}$ is taken into account, $y(t-2)$ evidently helps improve the prediction of $x(t)$. In this example, we have $z_{0,0,-1}(t) = \eta(t)$ such that the projection residuals of $x(t)$ and $y(t)$ onto $H\{z_{0,0,-1}(\infty)\}$ are given, respectively, by $\varepsilon(t)$ and $\xi(t)$. Thus, the $z_{0,0,-1}$-effect was eliminated, $\{x(t)\}$ and $\{y(t)\}$ remain orthogonal with each other, and spurious causality is not observed.

In view of these examples, the effect attributable to the third process $\{z(t)\}$ seems best eliminated by projecting onto neither $H\{z(\infty)\}$ nor $H\{z(t-1)\}$, but onto $H\{z_{0,0,-1}(\infty)\}$. Because the $z_{0,0,-1}(s)$ $(s > t)$ are orthogonal to $H\{x(t), y(t)\}$, the one-way effect elimination method does not accompany the difficulty of bringing the future z-effect in the present feedback relations between $x(t)$ and $y(t)$. In other words, the projections of $x(t)$ and $y(t)$ onto $H\{z_{0,0,-1}(\infty)\}$ are the same as the projections onto $H\{z_{0,0,-1}(t-1)\}$. Theorem 3.2 formally shows that the partial definition of this chapter does not accompany the spurious phenomenon defined by Hsiao (1982).

Sims' version of non-causality in the presence of the third series $\{z(t)\}$ is characterized as follows; see Sect. 2.3 and Sims (1972).

Theorem 3.1 $\{y(t)\}$ *does not partially cause* $\{x(t)\}$ *in the presence of* $\{z(t)\}$ *if and only if $y(t)$ is representable as*

$$y(t) = y^{(1)}(t) + y^{(2)}(t),$$

where $y^{(1)}(t)$ is the projection of $y(t)$ onto $H\{x(t), z_{0,0,-1}(t)\}$ and $y^{(2)}(t)$ is orthogonal to $H\{x(\infty), z_{0,0,-1}(\infty)\}$.

Proof Let $\bar{v}(t)$ and $v(t)$ be the projection and the residual of $y(t)$ onto $H\{z_{0,0,-1}(\infty)\}$, and let $\bar{u}(t)$ and $u(t)$ be the projection and the residual of $x(t)$ onto $H\{z_{0,0,-1}(\infty)\}$. According to Theorem 2.1, if $\{v(t)\}$ does not cause $\{u(t)\}$ in the Granger sense, $v(t)$ has the decomposition $v(t) = a(t) + b(t)$, where $a(t) \in H\{u(t)\}$ and $b(t) \in H\{u(\infty)\}^{\perp}$ for $H\{u(\infty)\}^{\perp}$ indicating the orthogonal complement of $H\{u(\infty)\}$ in H. Because $b(t) \in H\{z_{0,0,-1}(\infty)\}^{\perp}$, it follows that

$$b(t) \in H\{u(\infty), z_{0,0,-1}(\infty)\}^{\perp} = H\{x(\infty), z_{0,0,-1}(\infty)\}^{\perp}.$$

The necessity then follows from the relations

$$\bar{v}(t) + a(t) \in H\{u(t)\} \oplus H\{z_{0,0,-1}(t)\} = H\{x(t), z_{0,0,-1}(t)\}.$$

To prove sufficiency, consider the further decomposition of the assumed decomposition $y(t) = y^{(1)}(t) + y^{(2)}(t)$:

$$y^{(1)}(t) = v^{(1)}(t) + \bar{v}^{(1)}(t) \quad \text{and} \quad y^{(2)}(t) = v^{(2)}(t) + \bar{v}^{(2)}(t)$$

where $\bar{v}^{(1)}(t)$, $\bar{v}^{(2)}(t) \in H\{z_{0,0,-1}(\infty)\}$, $v^{(1)}(t)$, and $v^{(2)}(t)$ are the residuals. Then, by definition we have

$$v^{(1)}(t) \in H\{x(t)\} \cap H\{z_{0,0,-1}(\infty)\}^{\perp} = H\{u(t)\}$$

and

$$v^{(2)}(t) \perp \left\{ H\{u(\infty)\} \oplus H\{z_{0,0,-1}(\infty)\} \right\}.$$

Because $v(t) = v^{(1)}(t) + v^{(2)}(t)$, this completes the proof. $\qquad\square$

The next theorem generalizes the argument of Example 3.4. To discern the unconditional Granger causality limited on a pair of processes from the partial causality, the former is said simple causality.

Theorem 3.2 *If $\{y(t)\}$ does not simply cause $\{x(t)\}$, then $\{y(t)\}$ does not partially cause $\{x(t)\}$ in the presence of any third series $\{z(t)\}$.*

Proof Because $\{y(t)\}$ does not simply cause $\{x(t)\}$, $y(t)$ has the Sims representation $y(t) = y^{(1)}(t) + y^{(2)}(t)$ such that $y^{(1)}(t) \in H\{x(t)\} \subset H\{x(t), z_{0,0,-1}(t)\}$ and $y^{(2)}(t) \in H\{x(\infty)\}^{\perp}$. Let $y^{(3)}(t)$ be the projection of $y^{(2)}(t)$ onto $H\{x(t), z_{0,0,-1}(t)\}$ and $y^{(4)}(t)$ be the residual. Because $y^{(2)}(t) \in H\{x(\infty)\}^{\perp}$ and $y^{(2)}(t) \perp z_{0,0,-1}(t + j)$ for $j \in \mathbb{Z}^+$, it follows that $y^{(4)}(t) \in H\{x(\infty), z_{0,0,-1}(\infty)\}^{\perp}$. Consequently, setting $y^{(5)}(t) = y^{(1)}(t) + y^{(3)}(t)$, we have $y(t) = y^{(5)}(t) + y^{(4)}(t)$, where $y^{(5)}(t) \in H\{x(t), z_{0,0,-1}(t)\}$ and $y^{(4)}(t) \in H\{x(\infty), z_{0,0,-1}(\infty)\}^{\perp}$. The desired result then follows from Theorem 3.1. $\qquad\square$

3.3 Partial Measures of Interdependence

3.3.1 Representing the Partial Measures

For the three-series system $\{x(t), y(t), z(t)\}$, as defined in the previous section, the one-way effect component of $z(t)$ (which is denoted as $z_{0,0,-1}(t)$) is the projection residual of $z(t)$ onto the subspace generated by $\{x(s), y(s), z(s-1)\}$, $-\infty < s \le t$. Using the innovation orthogonalizing method of Sect. 2.4.2, the spectral density of the joint process $\{u(t), v(t)\}$, which is constructed from the process $\{x(t), y(t)\}$ by eliminating the $z_{0,0,-1}(t)$-effect, is derived from the spectral density $f(\lambda)$. First, the MA representation of the joint process $\{u(t), v(t)\}$ is derived as follows. Let $f(\lambda)$ be the joint spectral density of the non-deterministic full-rank second-order stationary process $w(t) = (x(t)^*, y(t)^*, z(t)^*)^*$, $t \in \mathbb{Z}$, and suppose that $f(\lambda)$ satisfies the Szegö condition

$$\int_{-\pi}^{\pi} \log \det f(\lambda)d\lambda > -\infty, \tag{3.1}$$

then the density has the canonical factorization

$$f(\lambda) = \frac{1}{2\pi} \Lambda(e^{-i\lambda})\Lambda(e^{-i\lambda})^* \tag{3.2}$$

by means of a $(p_1 + p_2 + p_3) \times (p_1 + p_2 + p_3)$ matrix $\Lambda(z)$, which is analytic and of full rank inside the unit disk. In (3.2), $\Lambda(e^{-i\lambda})$ is the boundary value of a maximal analytic function

$$\Lambda(z) = \sum_{j=0}^{\infty} \Lambda[j]z^j$$

with the real-matrix coefficients $\Lambda[j]$. A factorization such as (3.2) is said to be canonical. Let

$$\varepsilon(t) \equiv (\varepsilon_1(t)^*, \varepsilon_2(t)^*, \varepsilon_3(t)^*)^* \equiv w_{-1}(t)$$
$$\equiv (x_{-1,-1,-1}(t)^*, y_{-1,-1,-1}(t)^*, z_{-1,-1,-1}(t)^*)^*$$

be the one-step ahead prediction error of the process $w(t)$ by its past values. Denote the covariance matrix of $\varepsilon(t)$ by $\Sigma^\dagger$ and denote the partition matrix as

$$\Sigma^\dagger = \begin{bmatrix} \Sigma_{..}^\dagger & \Sigma_{.3}^\dagger \\ \Sigma_{3.}^\dagger & \Sigma_{33}^\dagger \end{bmatrix}.$$

Then, the residual of the projection of $\varepsilon_3(t)$ onto the linear space spanned by $\varepsilon_.(t) \equiv (\varepsilon_1(t)^*, \varepsilon_2(t)^*)^*$ is given by $\varepsilon_3(t) - \Sigma_{3.}^\dagger \Sigma_{..}^{\dagger^{-1}} \varepsilon_.(t)$, and it constitutes the one-way effect component of $z(t)$. For orthonormalizing the covariance matrix $Cov(\varepsilon_.(t), \varepsilon_3(t) - \Sigma_{3.}^\dagger \Sigma_{..}^{\dagger^{-1}} \varepsilon_.(t))$, define

$$\begin{bmatrix} \varepsilon_.^\dagger(t) \\ \varepsilon_3^\dagger(t) \end{bmatrix} = \begin{bmatrix} \Sigma_{..}^{\dagger^{-1/2}} & 0 \\ 0 & (\Sigma_{33..}^\dagger)^{-1/2} \end{bmatrix} \begin{bmatrix} I_{p_1+p_2} & 0 \\ -\Sigma_{3.}^\dagger \Sigma_{..}^{\dagger^{-1}} & I_{p_3} \end{bmatrix} \begin{bmatrix} \varepsilon_.(t) \\ \varepsilon_3(t) \end{bmatrix}, \tag{3.3}$$

and define a $(p_1 + p_2 + p_3) \times (p_1 + p_2 + p_3)$ matrix Δ by

$$\Delta \equiv \begin{bmatrix} \Sigma_{..}^{\dagger^{-1/2}} & 0 \\ 0 & (\Sigma_{33..}^\dagger)^{-1/2} \end{bmatrix} \begin{bmatrix} I_{p_1+p_2} & 0 \\ -\Sigma_{3.}^\dagger \Sigma_{..}^{\dagger^{-1}} & I_{p_3} \end{bmatrix},$$

where $\Sigma_{33..}^\dagger \equiv \Sigma_{33}^\dagger - \Sigma_{3.}^\dagger \Sigma_{..}^{\dagger^{-1}} \Sigma_{.3}^\dagger$. In view of the construction, Δ is a lower triangular block matrix

$$\Delta = \begin{bmatrix} \Delta_{..} & 0 \\ \Delta_{3.} & \Delta_{33} \end{bmatrix}.$$

Set $\tilde{\Lambda}(L) = \Lambda(L)\Lambda(0)^{-1}\Delta^{-1}$ and set its partition as

$$\tilde{\Lambda}(z) = \begin{bmatrix} \tilde{\Lambda}_{..}(z) & \tilde{\Lambda}_{.3}(z) \\ \tilde{\Lambda}_{3.}(z) & \tilde{\Lambda}_{33}(z) \end{bmatrix}.$$

Then, it follows from the relationships

$$w(t) \equiv (x(t)^*, y(t)^*, z(t)^*)^* = \Lambda(L)\Lambda(0)^{-1}\varepsilon(t) = \Lambda(L)\Lambda(0)^{-1}\Delta^{-1}\Delta\varepsilon(t)$$
$$\equiv \tilde{\Lambda}(L)\varepsilon^{\dagger}(t) \equiv \begin{bmatrix} \tilde{\Lambda}_{..}(L) & \tilde{\Lambda}_{.3}(L) \\ \tilde{\Lambda}_{3.}(L) & \tilde{\Lambda}_{33}(L) \end{bmatrix} \begin{bmatrix} \varepsilon_{.}^{\dagger}(t) \\ \varepsilon_3^{\dagger}(t) \end{bmatrix}$$

that

$$\begin{bmatrix} x(t) \\ y(t) \end{bmatrix} = \tilde{\Lambda}_{..}(L)\varepsilon_{.}^{\dagger}(t) + \tilde{\Lambda}_{.3}(L)\varepsilon_3^{\dagger}(t). \tag{3.4}$$

Because $\{\varepsilon_{.}^{\dagger}(t)\}$ and $\{\varepsilon_3^{\dagger}(t)\}$ are orthogonal, the spectral density of $\{x(t), y(t)\}$ is given by

$$f_{..}(\lambda) = \frac{1}{2\pi}\{\tilde{\Lambda}_{..}(e^{-i\lambda})\tilde{\Lambda}_{..}(e^{-i\lambda})^* + \tilde{\Lambda}_{.3}(e^{-i\lambda})\tilde{\Lambda}_{.3}(e^{-i\lambda})^*\}.$$

Then, in view of (3.4), $\{u(t), v(t)\}$ is represented by

$$\begin{bmatrix} u(t) \\ v(t) \end{bmatrix} = \tilde{\Lambda}_{..}(L)\varepsilon_{.}^{\dagger}(t), \tag{3.5}$$

whence the spectral density of $\{u(t), v(t)\}$ is represented by

$$h(\lambda) = \frac{1}{2\pi}\tilde{\Lambda}_{..}(e^{-i\lambda})\tilde{\Lambda}_{..}(e^{-i\lambda})^*. \tag{3.6}$$

In view of the construction, $\tilde{\Lambda}(z)$ is a canonical factor if $\Lambda(z)$ is canonical, but its square diagonal block $\tilde{\Lambda}_{..}(z)$ in (3.5) and (3.6) is not warranted to be so; see Remark 3.1. When the factor given in (3.6) is not canonical, a factorization procedure such as in Hosoya and Takimoto (2010) must be implemented because all partial measures proposed are constructed using the knowledge of a canonical factor of $h(\lambda)$.

The partial interdependency measures between $\{x(t)\}$ and $\{y(t)\}$ in the presence of $\{z(t)\}$ are defined as corresponding simple measures between $\{u(t)\}$ and $\{v(t)\}$ given in (3.5). It follows from the assumption that $f(\lambda)$ satisfies the Szegö condition that the process $\{u(t), v(t)\}$ is non-deterministic and $|\Lambda_{..}(e^{-i\lambda})| \neq 0$ a.e., since $|\Lambda(e^{-i\lambda})|$ is not equal to zero a.e.. Hence, the spectral density $h(\lambda)$ given in (3.6) satisfies the Szegö condition (3.1) and has a canonical factorization

$$h(\lambda) = \frac{1}{2\pi} \Gamma(e^{-i\lambda}) \Gamma(e^{-i\lambda})^*. \tag{3.7}$$

The relation (3.7) implies that the following time-domain MA representation of the series $\{u(t), v(t)\}$ holds in terms of the one-step ahead prediction error $\varepsilon(t) \equiv (\varepsilon_1(t)^*, \varepsilon_2(t)^*)^* \equiv (u_{-1,-1}(t)^*, v_{-1,-1}(t)^*)^*$; namely,

$$\begin{bmatrix} u(t) \\ v(t) \end{bmatrix} = \Gamma(L) \Gamma(0)^{-1} \begin{bmatrix} \varepsilon_1(t) \\ \varepsilon_2(t) \end{bmatrix},$$

where $E\{\varepsilon(t)\} = 0$ and $E\{\varepsilon(t)\varepsilon(t)^*\} = \Gamma(0)\Gamma(0)^* = \Sigma$. In parallel with (3.3),

$$\begin{bmatrix} \varepsilon_1^\dagger(t) \\ \varepsilon_2^\dagger(t) \end{bmatrix} \equiv \begin{bmatrix} \Sigma_{11}^{-1/2} & 0 \\ 0 & \Sigma_{22:1}^{-1/2} \end{bmatrix} \begin{bmatrix} I_{p_1} & 0 \\ -\Sigma_{21}\Sigma_{11}^{-1} & I_{p_2} \end{bmatrix} \begin{bmatrix} \varepsilon_1(t) \\ \varepsilon_2(t) \end{bmatrix} \tag{3.8}$$

$$\equiv \Xi\varepsilon(t),$$

whence $E\{\varepsilon^\dagger(t)\varepsilon^\dagger(t)^*\} = I_{p_1+p_2}$. Then, we have

$$\begin{bmatrix} u(t) \\ v(t) \end{bmatrix} = \Gamma(L)\Gamma(0)^{-1}\Xi^{-1}\Xi\varepsilon(t)$$

$$\equiv \Gamma^\dagger(L)\varepsilon^\dagger(t)$$

$$\equiv \begin{bmatrix} \Gamma_{11}^\dagger(L) & \Gamma_{12}^\dagger(L) \\ \Gamma_{21}^\dagger(L) & \Gamma_{22}^\dagger(L) \end{bmatrix} \begin{bmatrix} \varepsilon_1^\dagger(t) \\ \varepsilon_2^\dagger(t) \end{bmatrix}, \tag{3.9}$$

where $\{\varepsilon_2^\dagger(t)\}$ is the normalized one-way effect component of $v(t)$ to $u(t)$. The partial frequency-wise measure of one-way effect (FMO) from $\{y(t)\}$ to $\{x(t)\}$ in the presence of $\{z(t)\}$ is defined then by

$$PM_{y \to x:z}(\lambda) = \log \det\{I_{p_1} + \Gamma_{11}^\dagger(e^{-i\lambda})^{-1}\Gamma_{12}^\dagger(e^{-i\lambda})(\Gamma_{11}^\dagger(e^{-i\lambda})^{-1}\Gamma_{12}^\dagger(e^{-i\lambda}))^*\}, \tag{3.10}$$

where the $\Gamma_{ij}^\dagger(e^{-i\lambda})$ are defined in (3.9).

Remark 3.1 Suppose that a matrix $\tilde{\Lambda}(z) = \{\tilde{\Lambda}_{ij}(z), i, j = 1, 2, 3\}$ is given by

$$\tilde{\Lambda}(z) = \begin{bmatrix} 1 & 0 & 0 \\ 0 & 1 & 0 \\ 0 & 0 & d \end{bmatrix} - \begin{bmatrix} 1 & 0 & 0 \\ 0 & 2 & 1 \\ 0 & 1.001 & 0.5 \end{bmatrix} z;$$

then, all the zeros of $\det \tilde{\Lambda}(z)$ are either on or outside the unit circle if $-0.499 \le d \le -0.243$. In contrast, $\det \tilde{\Lambda}_{..}(z) = (1 - z)(1 - 2z)$ has one zero inside the unit circle, where $\tilde{\Lambda}_{..}(z)$ denotes the upper 2×2 diagonal block of $\tilde{\Lambda}(z)$. Consequently, when a partial spectral density is given as $h(\lambda)$ in (3.6), the factor $\tilde{\Lambda}_{\cdot\cdot}(e^{-i\lambda})$ on the right-hand side is not guaranteed to be canonical.

Remark 3.2 By extension of (3.3), we can carry out a block-wise orthogonalization of $\{\varepsilon_1(t), \varepsilon_2(t), \varepsilon_3(t)\}$ as

$$\begin{bmatrix} \ddot{\varepsilon}_1(t) \\ \ddot{\varepsilon}_2(t) \\ \ddot{\varepsilon}_3(t) \end{bmatrix} = \begin{bmatrix} I_{p_1} & 0 & 0 \\ -\Sigma_{21}^{\dagger}\Sigma_{11}^{\dagger-1} & I_{p_2} & 0 \\ -\Sigma_{3\cdot}^{\dagger}\Sigma_{\cdot\cdot}^{\dagger-1} & I_{p_3} \end{bmatrix} \begin{bmatrix} \varepsilon_1(t) \\ \varepsilon_2(t) \\ \varepsilon_3(t) \end{bmatrix}$$

$$\equiv J_{p_1+p_2+p_3}\varepsilon(t),$$

where $J_{p_1+p_2+p_3}$ is a block triangular matrix. Then, in terms of $\{\ddot{\varepsilon}_1(t), \ddot{\varepsilon}_2(t), \ddot{\varepsilon}_3(t)\}$, the series $\{w(t)\}$ can be represented as

$$w(t) \equiv \begin{bmatrix} x(t) \\ y(t) \\ z(t) \end{bmatrix} = \begin{bmatrix} M_{11}(L) & M_{12}(L) & M_{13}(L) \\ M_{21}(L) & M_{22}(L) & M_{23}(L) \\ M_{31}(L) & M_{32}(L) & M_{33}(L) \end{bmatrix} \begin{bmatrix} \ddot{\varepsilon}_1(t) \\ \ddot{\varepsilon}_2(t) \\ \ddot{\varepsilon}_3(t) \end{bmatrix} \qquad (3.11)$$

$$\equiv M(L)\ddot{\varepsilon}(t),$$

where the matrix $M(L)$ is defined by $M(L) = \Lambda(L)\Lambda(0)^{-1}J_{p_1+p_2+p_3}^{-1}$. This is the MA representation on which the "innovation accounting" of Sims (1980) is based. It follows from (3.11) that the $z(t)$-effect eliminated $\{x(t), y(t)\}$ process is given by

$$\begin{bmatrix} \ddot{x}(t) \\ \ddot{y}(t) \end{bmatrix} = \begin{bmatrix} M_{11}(L) & M_{12}(L) \\ M_{21}(L) & M_{22}(L) \end{bmatrix} \begin{bmatrix} \ddot{\varepsilon}_1(t) \\ \ddot{\varepsilon}_2(t) \end{bmatrix}. \qquad (3.12)$$

Because (3.11) is a canonical MA representation if $\Lambda(z)$ is a canonical factor, we may interpret that the component $M_{12}(L)\ddot{\varepsilon}_2(t) + M_{13}(L)\ddot{\varepsilon}_3(t)$ is the intrinsic effect attributable to $\{y(t), z(t)\}$. However, as noted in Remark 3.1, because the MA representation (3.12) is not necessarily canonical, the component $M_{12}(L)\ddot{\varepsilon}_2(t)$ is not guaranteed to be the intrinsic contribution attributable to $\{y(t)\}$ after the $z(t)$-effect elimination. Sim (1980) did not allude to this distinction. This consideration motivates the canonical factorization of this book.

Denote by $u'_{\cdot,\infty}(t)$ and $v'_{\infty,\cdot}(t)$, respectively, the projection residuals of $u(t)$, $v(t)$ onto $H\{v_{0,-1}(\infty)\}$, $H\{u_{-1,0}(\infty)\}$ and set their joint spectral density matrix as

$$h'(\lambda) = \begin{bmatrix} h'_{11}(\lambda) & h'_{12}(\lambda) \\ h'_{21}(\lambda) & h'_{22}(\lambda) \end{bmatrix}.$$

See Theorem 2.7 and (3.15) in the proof of the following theorem for the concrete representations of $h'(\lambda)$. Then, the partial measure of reciprocity at frequency λ between $\{x(t)\}$ and $\{y(t)\}$ in the presence of the third series $\{z(t)\}$ is defined by

$$PM_{x.y:z}(\lambda) \equiv M_{u.v}(\lambda) \equiv \log \left[\frac{\det h'_{11}(\lambda) \det h'_{22}(\lambda)}{\det h'(\lambda)} \right].$$

Set $\ddot{\sigma}^2 \equiv \det \Sigma_{11} \det \Sigma_{22} / \det \Sigma$.

Theorem 3.3 *Suppose that the spectral density matrix $\{u(t), v(t)\}$ has the canonical factorization (3.7). Then, we have*

$$PM_{x.y:z}(\lambda) = \log \ddot{\sigma}^2;$$

namely, the partial frequency-wise measure of reciprocity (FMR) is constant over the entire frequency domain. Hence, the overall measure $PM_{x.y:z}$ is equal to $\log \ddot{\sigma}^2$.

Proof It follows from the representation (3.9) that the reciprocal component $u'_{.,\infty}(t)$ of $u(t)$ is given by

$$u'_{.,\infty}(t) = \Gamma_{11}^\dagger(L) \Sigma_{11}^{-1/2} \varepsilon_1(t). \tag{3.13}$$

Similarly, setting

$$\Psi \equiv \begin{bmatrix} \Sigma_{11:2}^{-1/2} & 0 \\ 0 & \Sigma_{22}^{-1/2} \end{bmatrix} \begin{bmatrix} I_{p_1} & -\Sigma_{12}\Sigma_{22}^{-1} \\ 0 & I_{p_2} \end{bmatrix} \quad \text{and} \quad \xi(t) = \Psi \varepsilon(t),$$

we have

$$\begin{bmatrix} u(t) \\ v(t) \end{bmatrix} = \Gamma(L)\Gamma(0)^{-1}\Psi^{-1}\Psi \varepsilon(t)$$

$$\equiv \check{\Gamma}(L)\xi(t)$$

$$\equiv \begin{bmatrix} \check{\Gamma}_{11}(L) & \check{\Gamma}_{12}(L) \\ \check{\Gamma}_{21}(L) & \check{\Gamma}_{22}(L) \end{bmatrix} \begin{bmatrix} \xi_1(t) \\ \xi_2(t) \end{bmatrix}. \tag{3.14}$$

In view of the construction, $\{\xi_1(t)\}$ is the one-way effect component process of $\{u(t)\}$ to $\{v(t)\}$. It follows from the representations (3.13) and (3.14) that the reciprocal components of $u(t)$ and $v(t)$ are, respectively, given by

$$u'_{.,\infty}(t) = \Gamma_{11}^\dagger(L) \Sigma_{11}^{-1/2} \varepsilon_1(t) \quad \text{and} \quad v'_{\infty,.}(t) = \check{\Gamma}_{22}(L) \Sigma_{22}^{-1/2} \varepsilon_2(t).$$

Consequently, the joint spectral density matrix $h'(\lambda)$ of the process $\{u'_{.,\infty}(t), v'_{\infty,.}(t)\}$ is given by

$$
h'(\lambda) = \frac{1}{2\pi}
\begin{bmatrix}
\Gamma_{11}^{\dagger}(e^{-i\lambda})\Sigma_{11}^{-1/2} & 0 \\
0 & \check{\Gamma}_{22}(e^{-i\lambda})\Sigma_{22}^{-1/2}
\end{bmatrix}
\begin{bmatrix}
\Sigma_{11} & \Sigma_{12} \\
\Sigma_{21} & \Sigma_{22}
\end{bmatrix}
$$

$$
\times
\begin{bmatrix}
\{\Gamma_{11}^{\dagger}(e^{-i\lambda})\Sigma_{11}^{-1/2}\}^{*} & 0 \\
0 & \{\check{\Gamma}_{22}(e^{-i\lambda})\Sigma_{22}^{-1/2}\}^{*}
\end{bmatrix}
$$

$$
= \frac{1}{2\pi}
\begin{bmatrix}
\Gamma_{11}^{\dagger}(e^{-i\lambda})\Gamma_{11}^{\dagger}(e^{-i\lambda})^{*} & \Gamma_{11}^{\dagger}(e^{-i\lambda})\Sigma_{11}^{-1/2}\Sigma_{12}\Sigma_{22}^{-1/2}\check{\Gamma}_{22}(e^{-i\lambda})^{*} \\
\check{\Gamma}_{22}(e^{-i\lambda})\Sigma_{22}^{-1/2}\Sigma_{21}\Sigma_{11}^{-1/2}\Gamma_{11}^{\dagger}(e^{-i\lambda})^{*} & \check{\Gamma}_{22}(e^{-i\lambda})\check{\Gamma}_{22}(e^{-i\lambda})^{*}
\end{bmatrix}
$$

$$
= \frac{1}{2\pi}
\begin{bmatrix}
\Gamma_{11}^{\dagger}(e^{-i\lambda}) & 0 \\
0 & \check{\Gamma}_{22}(e^{-i\lambda})
\end{bmatrix}
\begin{bmatrix}
I_{p_1} & \Sigma_{11}^{-1/2}\Sigma_{12}\Sigma_{22}^{-1/2} \\
\Sigma_{22}^{-1/2}\Sigma_{21}\Sigma_{11}^{-1/2} & I_{p_2}
\end{bmatrix}
$$

$$
\times
\begin{bmatrix}
\Gamma_{11}^{\dagger}(e^{-i\lambda})^{*} & 0 \\
0 & \check{\Gamma}_{22}(e^{-i\lambda})^{*}
\end{bmatrix}.
\tag{3.15}
$$

It follows from (3.15) that

$$
\det h_{11}'(\lambda)\,\det h_{22}'(\lambda)/\det h'(\lambda) = \det\Sigma_{11}\,\det\Sigma_{22}/\det\Sigma \equiv \ddot{\sigma}^2. \qquad \square
$$

The quantity $\ddot{\sigma}^2$ defined in Theorem 3.3 is a constant not less than 1. Geweke (1982) termed the quantity $\ddot{\sigma}^2$ the measure of instantaneous feedback; see also Theorem 2.5. The following Theorem 3.4 has a useful application for the ARMA model, which enables representations of the measures of interdependence to be much simplified. Suppose that $\gamma(z)$ is a scalar-valued analytic function having an expansion with real coefficients defined on the complex plane such that $\gamma(0) = 1$ and that has no zeros inside the unit circle. Suppose then that the spectral density matrix $k(\lambda)$ of the process $\{u(t), v(t)\}$ is expressed in the form:

$$
k(\lambda) = |\gamma(e^{-i\lambda})|^2 h(\lambda),
\tag{3.16}
$$

Moreover, suppose that $h(\lambda) = (1/2\pi)\Gamma(e^{-i\lambda})\Gamma(e^{-i\lambda})^{*}$ for a canonical factor $\Gamma(z)$, such that we have a canonical factorization

$$
k(\lambda) = \frac{1}{2\pi}\gamma(e^{-i\lambda})\Gamma(e^{-i\lambda})\{\gamma(e^{-i\lambda})\Gamma(e^{-i\lambda})\}^{*}.
\tag{3.17}
$$

For this special case, we have the following result.

Theorem 3.4 *Suppose $\{u(t), v(t)\}$ has the spectral density $k(\lambda)$ given in (3.16). Then, the $M_{v\to u}(\lambda)$, $M_{u\to v}(\lambda)$, and $M_{u.v}(\lambda)$ are the same as the corresponding measures for the spectral density $h(\lambda)$.*

Proof Let $\varepsilon_i^{\dagger}(t)$ and $\Gamma_{ij}^{\dagger}(L), i, j = 1, 2$ be defined as in (3.9) based on the factorization $h(\lambda) = \frac{1}{2\pi}\Gamma(e^{-i\lambda})\Gamma(e^{-i\lambda})^{*}$. If the spectral density $k(\lambda)$ has the canonical factorization (3.17), parallel to (3.8), we have the time-domain representation

$$\begin{bmatrix} u(t) \\ v(t) \end{bmatrix} = \gamma(L)\Gamma(L)\Gamma(0)^{-1} \begin{bmatrix} \varepsilon_1(t) \\ \varepsilon_2(t) \end{bmatrix}$$

$$= \begin{bmatrix} \Gamma_{11}^{\dagger\dagger}(L) & \Gamma_{12}^{\dagger\dagger}(L) \\ \Gamma_{21}^{\dagger\dagger}(L) & \Gamma_{22}^{\dagger\dagger}(L) \end{bmatrix} \begin{bmatrix} \varepsilon_1^{\dagger}(t) \\ \varepsilon_2^{\dagger}(t) \end{bmatrix} \qquad (3.18)$$

where

$$\Gamma_{ij}^{\dagger\dagger} = \gamma(L)\Gamma_{ij}^{\dagger}(L), i = 1, 2.$$

Hence, we have

$$M_{v \to u}(\lambda) \equiv \log \frac{\det\{\Gamma_{11}^{\dagger\dagger}(e^{-i\lambda})\Gamma_{11}^{\dagger\dagger}(e^{-i\lambda})^* + \Gamma_{12}^{\dagger\dagger}(e^{-i\lambda})\Gamma_{12}^{\dagger\dagger}(e^{-i\lambda})^*\}}{\det\{\Gamma_{11}^{\dagger\dagger}(e^{-i\lambda})\Gamma_{11}^{\dagger\dagger}(e^{-i\lambda})^*\}}$$

$$= \log \frac{\det\{\Gamma_{11}^{\dagger}(e^{-i\lambda})\Gamma_{11}^{\dagger}(e^{-i\lambda})^* + \Gamma_{12}^{\dagger}(e^{-i\lambda})\Gamma_{12}^{\dagger}(e^{-i\lambda})^*\}}{\det\{\Gamma_{11}^{\dagger}(e^{-i\lambda})\Gamma_{11}^{\dagger}(e^{-i\lambda})^*\}}. \qquad (3.19)$$

Namely, the right-hand-side member of (3.19) implies that the FMO based on $k(\lambda)$ is equal to the FMO for the spectral density $h(\lambda)$. In the same manner, for the process given by (3.18), the joint spectral density matrix $\breve{k}(\lambda)$ of the reciprocal component process $\{u'_{.,\infty}(t), v'_{\infty,.}(t)\}$ is equal to $|\gamma(e^{-i\lambda})|^2 h'(\lambda)$, where $h'(\lambda)$ is the density given by (3.15). Therefore, the frequency-wise measure of reciprocity is given by

$$M_{u.v}(\lambda) = \log[\det\{|\gamma(e^{-i\lambda})|^2 h'_{11}(\lambda)\} \det\{|\gamma(e^{-i\lambda})|^2 h'_{22}(\lambda)\} / \det\{|\gamma(e^{-i\lambda})|^2 h'(\lambda)\}$$

$$= \log \ddot{\sigma}^2. \qquad \square$$

Remark 3.3 Breitung and Candelon (2006, p. 364) directly derived $\varepsilon^{\dagger}(t)$ in (3.8) by multiplying the Cholesky factor matrix of the inverse of the covariance matrix of $\varepsilon(t)$. In the case of their bivariate model in which $u(t)$ and $v(t)$ are scalar-valued, if the orthogonalization is done by the lower triangle Cholesky matrix, the one-way effect component is automatically derived because then orthogonalization is conducted by eliminating the effect of $\varepsilon_2(t)$ from $\varepsilon_1(t)$ via the projection. In general, however, when $u(t)$ and $v(t)$ are vector-valued, arbitrary orthogonalization of $\varepsilon_1(t)$ and $\varepsilon_2(t)$ does not necessarily produce the one-way effect measure.

3.3.2 Glossary on Partial Measures of Interdependence

This subsection collects basic definitions and equations related to partial interdependencies. In the system of three series $\{x(t), y(t), z(t)\}$, the one-way effect component of $z(t)$ implies the projection residual (the perpendicular) of $z(t)$ when it is projected onto the closed linear subspace $H\{x(t), y(t), z(t-1)\}$ and the residual is denoted by $z_{0,0,-1}(t)$. The vectors $u(t)$ and $v(t)$ are, respectively, the projection residuals of $x(t)$ and $y(t)$ onto $H\{z_{0,0,-1}(t-j); j \in \mathbb{Z}\}$.

D1 The partial overall measure of one-way effect (OMO) from $\{y(t)\}$ to $\{x(t)\}$ is the simple OMO from $\{v(t)\}$ to $\{u(t)\}$ and is defined by

$$PM_{y \to x:z} \equiv M_{v \to u} = \log \frac{\det Cov\{u_{-1,.}(t)\}}{\det Cov\{u'_{-1,-1}(t)\}},$$

where $u_{-1,.}(t)$ and $u'_{-1,-1}(t)$ are the projection residuals of $u(t)$ onto $H\{u(t-1)\}$ and onto $H\{u(t-1), v_{0,-1}(t-1)\}$, respectively.

D2 The partial frequency-wise measure of one-way effect (FMO) is defined by

$$PM_{y \to x:z}(\lambda) \equiv M_{v \to u}(\lambda)$$
$$= \log \frac{\det\{\Gamma_{11}^\dagger(e^{-i\lambda})\Gamma_{11}^\dagger(e^{-i\lambda})^* + \Gamma_{12}^\dagger(e^{-i\lambda})\Gamma_{12}^\dagger(e^{-i\lambda})^*\}}{\det\{\Gamma_{11}^\dagger(e^{-i\lambda})\Gamma_{11}^\dagger(e^{-i\lambda})^*\}}$$

where the $\Gamma_{ij}^\dagger(e^{-i\lambda})$ are defined in (3.9). The partial FMO $PM_{x \to y:z}(\lambda)$ is given in a similar manner; see Chap. 2 for different representations.

D3 The partial measure of reciprocity at frequency λ and the corresponding overall measure between $x(t)$ and $y(t)$ are defined, respectively, by:

$$PM_{x.y:z}(\lambda) \equiv M_{u.v}(\lambda) = \log\left[\frac{\det h'_{11}(\lambda) \det h'_{22}(\lambda)}{\det h'(\lambda)}\right] = \log \ddot{\sigma}^2;$$

$$PM_{x.y:z} \equiv \frac{1}{2\pi}\int_{-\pi}^{\pi} PM_{x.y:x}(\lambda)d\lambda = \log \ddot{\sigma}^2.$$

D4 The partial measure of association at frequency λ and the corresponding overall measure between $x(t)$ and $y(t)$ are defined, respectively, by:

$$PM_{x,y:z}(\lambda) \equiv M_{u,v}(\lambda) \equiv \log\left[\frac{\det h_{11}(\lambda) \det h_{22}(\lambda)}{\det h'(\lambda)}\right],$$

$$PM_{x,y:z} \equiv \frac{1}{2\pi}\int_{-\pi}^{\pi} PM_{x,y:z}(\lambda)d\lambda.$$

where $h(\lambda)$ is the spectral density matrix of the joint process $\{u(t), v(t)\}$.

E1 The following equality holds between the partial OMO and FMO:

$$PM_{y \to x:z} = \frac{1}{2\pi}\int_{-\pi}^{\pi} PM_{y \to x:z}(\lambda)d\lambda;$$

see Theorem 2.2 for the proof.

E2 It follows from the definitions of the respective measures, the equality E1, and the corresponding equality for $PM_{x \to y:z}$ that:

$$PM_{x,y:z}(\lambda) = PM_{x \to y:z}(\lambda) + PM_{x.y:z}(\lambda) + PM_{y \to x:z}(\lambda),$$
$$PM_{x,y:z} = PM_{x \to y:z} + PM_{x.y:z} + PM_{y \to x:z}.$$

3.3.3 The Stationary ARMA Model

Suppose that the process $\{x(t), y(t), z(t)\}$ is a stationary multivariate ARMA process generated by

$$A(L) \begin{bmatrix} x(t) \\ y(t) \\ z(t) \end{bmatrix} = B(L)\varepsilon(t), \quad t \in \mathbb{Z}, \tag{3.20}$$

where $x(t), y(t), z(t)$ are, respectively, p_1, p_2, p_3 vectors, $A(L)$ and $B(L)$ are ath and bth order polynomials of the lag operator L, namely $A[0] = B[0] = I_{p_1+p_2+p_3}$ and $A(L) = \sum_{j=0}^{a} A[j]L^j$ and $B(L) = \sum_{j=0}^{b} B[j]L^j$.

We assume that all zeros of $\det A(z)$ are outside the unit circle, and $\det B(z)$ has zeros either on or outside the unit circle and does not share any common zeros with $\det A(z)$. Moreover, we assume that the innovation $\{\varepsilon(t)\}$ is a white noise process with mean 0 and covariance matrix $\Sigma^\dagger$. Because of the zero conditions of $A(z)$ and $B(z)$, the joint spectral density $f(\lambda)$ of the process (3.20) satisfies the Szegö condition (3.1), whence it has a canonical factorization

$$f(\lambda) = \frac{1}{2\pi} \Lambda(e^{-i\lambda}) \Lambda(e^{-i\lambda})^*.$$

In view of the zero conditions of $A(z)$ and $B(z)$, a version of the canonical factor $\Lambda(z)$ is given by

$$\Lambda(z) = A(z)^{-1} B(z) \Sigma^{\dagger \frac{1}{2}} = (\det A(z))^{-1} A^\sharp(z) B(z) \Sigma^{\dagger \frac{1}{2}} \equiv (\det A(z))^{-1} C(z),$$

where $\Sigma^{\dagger \frac{1}{2}}$ is the Cholesky factor of $\Sigma^\dagger$ satisfying $\Sigma^\dagger = \Sigma^{\dagger \frac{1}{2}} (\Sigma^{\dagger \frac{1}{2}})^*$; $A^\sharp(z)$ denotes the adjugate matrix (transposed cofactor matrix) of $A(z)$ and $C(z) (\equiv A^\sharp(z) B(z) \Sigma^{\dagger \frac{1}{2}})$ is a finite-order real-matrix coefficient polynomial such that

$$C(z) = \sum_{j=0}^{\bar{a}} C[j]z^j, \quad \bar{a} \equiv (p_1 + p_2 + p_3 - 1)a + b.$$

Denote the projection residuals of $x(t)$ and $y(t)$ onto $H\{z_{0,0,-1}(\infty)\}$, respectively, by $u(t)$ and $v(t)$, and denote the joint spectral density matrix of $\{u(t), v(t)\}$ by $h(\lambda)$. Now, set

$$
\tilde{\Lambda}(z) = C(z) \begin{bmatrix} \Sigma_{\cdot\cdot}^{\dagger-1/2} & 0 \\ 0 & (\Sigma_{33\cdot}^{\dagger})^{-1/2} \end{bmatrix} \begin{bmatrix} I_{p_1+p_2} & 0 \\ -\Sigma_{3\cdot}^{\dagger}\Sigma_{\cdot\cdot}^{\dagger-1} & I_{p_3} \end{bmatrix}
$$

and let $\tilde{\Lambda}_{\cdot\cdot}(z)$ be the $(p_1 + p_2) \times (p_1 + p_2)$ upper diagonal block of $\tilde{\Lambda}(z)$. Parallel to (3.6), the spectral density $h(\lambda)$ of $\{u(t), v(t)\}$ is given by

$$
h(\lambda) = \frac{1}{2\pi} |\det A(e^{-i\lambda})|^{-2} \tilde{\Lambda}_{\cdot\cdot}(e^{-i\lambda}) \tilde{\Lambda}_{\cdot\cdot}(e^{-i\lambda})^*.
$$

In view of Theorem 3.4, all interdependency measures between $\{u(t)\}$ and $\{v(t)\}$ are able to be constructed from the spectral density $k(\lambda) = \frac{1}{2\pi} \tilde{\Lambda}_{\cdot\cdot}(e^{-i\lambda}) \tilde{\Lambda}_{\cdot\cdot}(e^{-i\lambda})^*$. Since the zeros of $C(z)$ are equal to those of $\Lambda(z)$ in view of the construction, none of them are inside of the unit circle. Hence, $\tilde{\Lambda}(e^{-i\lambda})$ is a canonical factor of $k(\lambda)$. Consequently, all measures of interdependence exhibited in Sect. 3.3.2 can be computed using the factor $\tilde{\Lambda}(z)$.

3.4 Extension to Non-stationary Reproducible Processes

There exists a class of possibly non-stationary processes to which the prediction theory of stationary processes naturally extends, and they are termed reproducible processes. Let $\{w(t), \ t \in \mathbb{Z}\}$ be a second-order stationary p-vector process, and let $\{W(t), \ t \in \mathbb{Z}^+\}$ be another p-vector process with finite second-order moments. In the following arguments, the convention $W(t) = 0$ for $t \in \mathbb{Z}^{0-}$ is used whenever $W(t)$ is originally defined only for $t \in \mathbb{Z}^+$ and the extension is needed. Let $w_{-1}(t)$ be the one-step ahead prediction error of $w(t)$ based on $H\{w(t-1)\}$; then, the process $\{W(t)\}$ is said to be reproducible by (the generating process) $\{w(t)\}$ if the projection residual of $W(t)$ onto $H\{W(t-1), w(0)\}$ is equal to $w_{-1}(t)$ for all $t \in \mathbb{Z}^+$. Namely, the respective prediction errors of $w(t)$ and $W(t)$ are identical if the information $H\{w(0)\}$ is supplemented. If a process $\{W(t)\}$ is reproducible by $\{w(t)\}$, then it evidently follows that $H\{W(t), w(0)\} = H\{w(t)\}$ for $t \in \mathbb{Z}^+$.

Example 3.5 Let $A(L, t) = \sum_{j=0}^{a} A(j, t)L^j$ be a finite-order time-dependent linear filter, the $A(j, t)$s are $p \times p$ matrices, and $A(0, t) = I_p$ for all $t \in \mathbb{Z}^+$. Given a second-order stationary process $\{w(t)\}$, let $\{W(t), \ t \in \mathbb{Z}^+\}$ be generated by $A(L, t)W(t) = w(t)$. Then, the process $\{W(t)\}$ is reproducible by $\{w(t)\}$.

The Granger causality is defined between a pair of reproducible processes, as follows.

Definition 3.1 Suppose that the processes $\{U(t)\}$ and $\{V(t)\}$ are reproducible with respect to $\{u(t)\}$ and $\{v(t)\}$, respectively. If the covariance matrices of the prediction error of $U(t)$ based, respectively, on $H\{u(t-1),\ v(t-1)\}$, and $H\{u(t-1)\}$ are identical, $\{V(t)\}$ is said not to cause $\{U(t)\}$.

Then, the next theorem is a straightforward consequence of this extended definition of Granger non-causality.

Theorem 3.5 $\{V(t)\}$ *does not cause* $\{U(t)\}$ *if and only if* $\{v(t)\}$ *does not cause* $\{u(t)\}$.

Let $\{\varepsilon(t),\ t \in \mathbb{Z}\}$ be a p-vector white noise process with mean 0 and covariance matrix Σ and suppose that $\{W(t),\ t \geq 1\}$ is a p-vector linear process with time-varying coefficients generated by

$$W(t) = \sum_{j=0}^{\infty} A[j,\ t]\varepsilon(t-j), \tag{3.21}$$

where $A[0,\ t] = I_p$ and $\sum_{j=0}^{\infty} A[j,\ t]\Sigma A[j,\ t]^* < \infty$ for all $t \in \mathbb{Z}^+$; hence, $W(t)$ has a finite covariance matrix for each t. Suppose also that the process $\{W(t),\ t \in \mathbb{Z}^+\}$ in (3.21) is a second-order process reproducible by $\{\varepsilon(t)\}$; namely, $H\{W(t),\ \varepsilon(0)\} = H\{\varepsilon(t)\}$ for all $t \in \mathbb{Z}^+$. Now, set

$$A(e^{-i\lambda},\ t) = \sum_{j=0}^{\infty} A(j,t)e^{-i\lambda j}$$

and define the evolutionary spectral density matrix $f^W(\lambda,\ t)$ $(t \in \mathbb{Z}^+)$ of the process $\{W(t)\}$ by

$$f^W(\lambda,\ t) = \frac{1}{2\pi} A(e^{-i\lambda},\ t)\Sigma A(e^{-i\lambda},\ t)^*, \quad -\pi < \lambda \leq \pi,$$

(see Priestley (1988) for a detailed discussion of evolutionary spectra). The next theorem asserts that the prediction error formula (2.7) is extended to this class of reproducible linear processes; the proof is found in Hosoya (1997a).

Theorem 3.6 *If a reproducible process* $\{W(t)\}$ *in (3.21) has time-varying density matrix* f^W *such that for any* $t \in \mathbb{Z}^+$, *the Szegö condition*

$$\int_{-\pi}^{\pi} \log \det\ f^W(\lambda,\ t)d\lambda > -\infty;$$

holds. Then, for $t \in \mathbb{Z}^+$ *we have*

$$\det Cov\{W_{-1}(t)\} = (2\pi)^p \exp\{\frac{1}{2\pi} \int_{-\pi}^{\pi} \log \det\ f^W(\lambda,\ t)d\lambda\},$$

where $W_{-1}(t)$ is the projection residual of $W(t)$ onto $H\{W(t-1), \varepsilon(0)\}$.

The partial measures of Sect. 3.3 can be extended to the time-varying coefficient process (3.21) when $\{W(t)\}$ is constituted as $W(t) = (X(t)^*, Y(t)^*, Z(t)^*)$ and if the processes $X(t)$, $Y(t)$, and $Z(t)$ are individually reproducible with respect to appropriate generating processes.

Let $d(L) = \sum_{j=0}^{s} d[j]L^j$ be a lag operator with scalar coefficients such that $d(0) = 1$ and the zeros of $\sum_{j=0}^{s} d[j]z^j$ are either on or outside the unit circle, and let $D(L)$ be the $(p_1 + p_2 + p_3) \times (p_1 + p_2 + p_3)$ diagonal matrix that has the common diagonal element $d(L)$. Now, consider the possibly non-stationary processes that are represented by

$$D(L) \begin{bmatrix} X(t) \\ Y(t) \\ Z(t) \end{bmatrix} = \begin{bmatrix} x(t) \\ y(t) \\ z(t) \end{bmatrix} \tag{3.22}$$

where $\{x(t), y(t), z(t)\}$ is a second-order stationary process. Note that the processes $\{X(t)\}$, $\{Y(t)\}$, and $\{Z(t)\}$ are individually reproducible with respect to $\{x(t)\}$, $\{y(t)\}$, and $\{z(t)\}$, respectively. For the model (3.22), the causal concepts and measures are extended as follows. Define the one-way effect component of $Z(t)$ as the projection residual of $Z(t)$ onto $H\{X(t), Y(t), Z(t-1), x(0), y(0), z(0)\}$; thus, this is equal to $z_{0,0,-1}(t)$. The projection residuals of $X(t)$ onto $H\{X(t-1), x(0), z_{0,0,-1}(\infty)\}$ and $H\{X(t-1), Y(t-1), x(0), y(0), z_{0,0,-1}(\infty)\}$ are equal to the ones of $x(t)$ onto $H\{x(t-1), z_{0,0,-1}(\infty)\}$ and $H\{x(t-1), y(t-1), z_{0,0,-1}(\infty)\}$, respectively. As in Sect. 3.3, if we denote the projection residuals of $x(t)$ and $y(t)$ onto $H\{z_{0,0,-1}(\infty)\}$ by $u(t)$ and $v(t)$, the partial measure of one-way effect from $\{Y(t)\}$ to $\{X(t)\}$ is determined by the measure of one-way effect from $\{v(t)\}$ and $\{u(t)\}$. Therefore, it is natural to define extendedly the partial measures of the one-way effect from $\{Y(t)\}$ to $\{X(t)\}$ in the presence of $\{Z(t)\}$ by the corresponding measures $PM_{Y \rightarrow X:Z}$ and $PM_{Y \rightarrow X:Z}(\lambda)$, as given in Sect. 3.3.2; namely, $PM_{Y \rightarrow X:Z} = PM_{y \rightarrow x:z}$ and $PM_{Y \rightarrow X:Z}(\lambda) = PM_{y \rightarrow x:z}$. The measures $PM_{X \rightarrow Y:Z}$ and $PM_{X \rightarrow Y:Z}(\lambda)$ are defined analogously. Moreover, the measures of association and reciprocity can be defined in a parallel manner for a possibly non-stationary process $\{X(t), Y(t), Z(t)\}$.

Suppose that $\{X(t)\}$, $\{Y(t)\}$, $\{Z(t)\}$, respectively, are p_1, p_2, p_3-vector processes and set $W(t) = (X(t)^*, Y(t)^*, Z(t)^*)^*$. Such a $(p_1 + p_2 + p_3)$-vector process $W(t)$ as (3.22) is typically generated by a cointegrated ARMA process. Set

$$A(L) = \sum_{j=0}^{a} A[j]L^j \text{ and } B(L) = \sum_{k=0}^{b} B[k]L^k$$

where the $A[j]$ and the $B[j]$ are $(p_1 + p_2 + p_3) \times (p_1 + p_2 + p_3)$ matrices with $A[0] = B[0] = I_{p_1+p_2+p_3}$. We assume that the zeros of $\det A(z)$ and $\det B(z)$ are either on or outside the unit circle, and $\det A(z)$ and $\det B(z)$ do not share common zeros. Suppose $\{W(t)\}$ is generated by

$$A(L)W(t) = B(L)\varepsilon(t) \tag{3.23}$$

for a white noise process $\{\varepsilon(t)\}$ such that $E(\varepsilon(t)) = 0$ and $Cov(\varepsilon(t)) = \Sigma^{\dagger}$, where $rank(\Sigma^{\dagger}) = p_1 + p_2 + p_3$. Denote then by $A(L)^{\sharp}$ the adjugate matrix of $A(L)$ such that

$$A(L)^{\sharp} A(L) = \begin{bmatrix} d(L) & & 0 \\ & \ddots & \\ 0 & & d(L) \end{bmatrix} \equiv D(L)$$

where $D(L)$ is the diagonal matrix with $\det A(L)$ as the common diagonal element. Applying the operator $A(L)^{\sharp}$ to the members of the Eq. (3.23), we have

$$\begin{bmatrix} d(L) & & 0 \\ & \ddots & \\ 0 & & d(L) \end{bmatrix} \begin{bmatrix} X(t) \\ Y(t) \\ Z(t) \end{bmatrix} = A(L)^{\sharp} B(L)\varepsilon(t)$$

$$\equiv \begin{bmatrix} x(t) \\ y(t) \\ z(t) \end{bmatrix} \equiv w(t). \tag{3.24}$$

It follows from the construction that $\{w(t)\}$ is a stationary MA process; moreover, because the zeros of $\det\{A(z)^{\sharp} B(z)\}$ are either on or outside the unit circle, the covariance matrix of the one-step ahead prediction error of $w(t)$ is equal to $\Sigma^{\dagger}$ and the joint spectral density matrix $f(\lambda)$ of $\{w(t)\}$ has canonical factorization

$$f(\lambda) = \frac{1}{2\pi} \Lambda(e^{-i\lambda}) \Lambda(e^{-i\lambda})^{*},$$

where $\Lambda(e^{-i\lambda}) = A(e^{-i\lambda})^{\sharp} B(e^{-i\lambda}) \Sigma^{\dagger 1/2}$, where $\Sigma^{\dagger 1/2} \Sigma^{\dagger 1/2} = \Sigma^{\dagger}$. Following this extended definition, the partial structure of interdependence in the cointegrated process in (3.23) is identified with that in $\{x(t)\}$, $\{y(t)\}$, and $\{z(t)\}$, and all partial measures of the possibly non-stationary processes $\{X(t)\}, \{Y(t)\}, \{Z(t)\}$ can be constructed through the corresponding measures for the stationary processes $\{x(t)\}$, $\{y(t)\}$, and $\{z(t)\}$.

For the cointegrated AR model, Hosoya (1997a) and Yao and Hosoya (2000) applied the Wald test approach for the inference of the simple one-way effect measures; see Johansen (1995) for an allied asymptotic theory of the reduced rank maximum-likelihood estimator for cointegrated VAR processes.

Remark 3.4 Granger and Lin (1995) presented an analytic characterization of long-run causality for a bivariate cointegrated AR model using the one-way effect measure. The reduction (3.24) to derive the one-way effect measure of this section is a generalization of their idea. Suppose that a two-variated cointegration model is given by

$$\Delta x(t) = \lambda_1(y(t) - cx(t)) + a_1(L)\Delta x(t-1) + b_1(L)\Delta y(t-1) + \varepsilon_1(t)$$
$$\Delta y(t) = \lambda_2(y(t) - cx(t)) + a_2(L)\Delta x(t-1) + b_2(L)\Delta y(t-1) + \varepsilon_2(t) \quad (3.25)$$

where $\varepsilon_1(t)$ and $\varepsilon_2(t)$ are white noise processes with mean 0 and variance 1 and $Cov(\varepsilon_1(t), \varepsilon_2(t)) = \rho$; $a(L)$ and $b(L)$ are possibly infinite-order polynomials of L. For the cointegration model (3.25), they provide the one-way effect measure $M_{y \to x}(\omega)$ in (13) of Granger and Lin (1995, p. 534). In the paper, they assert based on that measure that there is no long-run one-way causality among full-rank unit-root VAR processes. But the assertion does not seem to hold in general. Set $\lambda_1 = \lambda_2 = 0$ in (3.25) such that there is no cointegration; then, the one-way effect measure $y \to x$ becomes in their notations

$$M_{y \to x}(\omega) = \log[1 + b_1^2(1 - \rho^2)/|\bar{z} - b_2 + \rho b_1|^2],$$

where $\bar{z} = e^{-i\omega}$, $b_1 = b_1(e^{i\omega})$ and $b_2 = b_2(e^{i\omega})$. Hence, if $|\omega| \to 0$ and $|\bar{z}| \to 1$, we have

$$\lim_{|\omega| \to 0} M_{y \to x}(\omega) = \log[1 + b_1^2(1 - \rho^2)/(1 - b_2 + \rho b_1)^2]$$

where the right-hand-side member is not necessarily equal to 0.

References

Breitung, J., & Candelon, B. (2006). Testing for short- and long-run causality: A frequency-domain approach. *Journal of Econometrics, 132*, 363–378.

Geweke, J. (1982). Measurement of linear dependence and feedback between multiple time series. *Journal of the American Statistical Association, 77*, 304–324.

Geweke, J. (1984). Measures of conditional linear dependence and feedback between time series. *Journal of the American Statistical Association, 79*, 907–915.

Granger, C. W. J. (1969). Investigating causal relations by econometric methods and cross-spectral methods. *Econometrica, 37*, 424–438.

Granger, C. W. J., & Lin, J. L. (1995). Causality in the long run. *Econometric Theory, 11*, 530–536.

Hosoya, Y. (1991). The decomposition and measurement of the interdependency between second-order stationary processes. *Probability Theory and Related Fields, 88*, 429–444.

Hosoya, Y. (1997a). Causal analysis and statistical inference on possibly non-stationary time series. In D.M. Kreps & K.F. Wallis (Eds.), *Advances in Economics and Econometrics: Theory and Application, Seventh World Congress* (Vol. III, Chap. 1, pp. 1–33). Cambridge: Cambridge University press.

Hosoya, Y. (2001). Elimination of third-series effect and defining partial measures of causality. *Journal of Time Series Analysis, 22*, 537–554.

Hosoya, T., & Takimoto, T. (2010). A numerical method for factorizing the rational spectral density matrix. *Journal of Time Series Analysis, 31*, 229–240.

Hsiao, C. (1982). Time series modelling and causal ordering of Canadian money, income and interest rates. In O. D. Anderson (Ed.), *Time Series Analysis: Theory and Practice I* (pp. 671–698). Amsterdam: North-Holland.

Johansen, S. (1995). *Likelihood-based Inference in Cointegrated Autoregressive Models*. Oxford: Oxford University Press.

Priestley, M. B. (1988). *Non-linear and Non-stationary Time Series Analysis*. London: Academic Press.

Sims, C. A. (1972). Money, income and causality. *American Economic Review*, *62*, 540–552.

Sims, C.A. (1980). Comparison of interwar and postwar business cycles: Monetarism reconsidered. *The American Economic Review*, *70*, 250–257.

Yao, F., & Hosoya, Y. (2000). Inference on one-way effect and evidence in Japanese macroeconomic data. *Journal of Econometrics*, *98*, 225–255.

Chapter 4
Inference Based on the Vector Autoregressive and Moving Average Model

Abstract Based on the stationary vector ARMA process, this chapter shows how the partial measures of interdependence introduced in Sect. 3.3 are numerically evaluated and applied to practical situations. Section 4.1 discusses the statistical inference on those measures using the standard asymptotic theory of the Whittle likelihood inference for stationary multivariate ARMA processes. The point is the use of simulation-based estimations of the covariance matrix of each measure-related statistic. In Sect. 4.2, we investigate the small sample performance of partial one-way effect measure estimates using Monte Carlo data generated by a pair of trivariate data generating processes, the VAR(2) and VARMA(1,1) models. All model parameter estimates are produced using an improved version of the Takimoto and Hosoya (2004, 2006) procedure. The partial frequency-wise measures of the one-way effect are evaluated using spectral factorization, and the parameters are substituted with a modified Whittle estimate. To illustrate the analysis of interdependence in the frequency domain, Sect. 4.3 provides an empirical analysis of US interest rates and economic growth data.

Keywords Inference procedure · Monte Carlo Wald test · Partial measures of interdependence · Small sample performance · US macroeconomic data · Vector ARMA process

4.1 Inference Procedure

This section discusses the statistical inference of the partial measures of interdependence introduced in Sect. 3.3 using the standard asymptotic theory of the Whittle likelihood inference for multivariate stationary ARMA processes. For testing purposes, we use a simulation-based estimation of the covariance matrix of each measure-related statistic.

© The Author(s) 2017   65
Y. Hosoya et al., *Characterizing Interdependencies of Multiple Time Series*,
JSS Research Series in Statistics, DOI 10.1007/978-981-10-6436-4_4

4.1.1 Three-Step Estimation Procedure

As in Sect. 3.3.3, suppose that $\{x(t), y(t), z(t)\}$ is a stationary multivariate ARMA process generated by

$$A(L)\begin{bmatrix} x(t) \\ y(t) \\ z(t) \end{bmatrix} = B(L)\varepsilon(t), \quad t \in \mathbb{Z}, \tag{4.1}$$

where $x(t)$, $y(t)$, and $z(t)$ are, respectively, p_1, p_2, and p_3 vectors, $A(L)$ and $B(L)$ are ath- and bth-order polynomials of the lag operator L, and $A[0] = B[0] = I_{p_1+p_2+p_3}$. Namely, we have $A(L) = \sum_{j=0}^{a} A[j]L^j$ and $B(L) = \sum_{j=0}^{b} B[j]L^j$, and the sizes of the coefficient matrices $A[j]$ and $B[j]$ are $(p_1 + p_2 + p_3) \times (p_1 + p_2 + p_3)$. The assumptions of Sect. 3.3.3 on the process (4.1) are retained.

Based on a finite set of observations $\{x(t), y(t), z(t); t = 1, \cdots, T\}$ and the VARMA model (4.1) for the data generating process, we are able to conduct statistical inference on the partial measures of interdependence introduced in Sect. 3.3. Let θ be a n_θ-vector parameter and assume that the model parameter of (4.1) is a function of θ. Namely,

$$vec\{A, B, v(\Sigma^\dagger)\} = vec\{A(\theta), B(\theta), v(\Sigma^\dagger)(\theta)\},$$

where $A = vec\{A[1], \cdots, A[a]\}$, $B = vec\{B[1], \cdots, B[b]\}$, and $v(\Sigma^\dagger)$ denotes the $(p_1 + p_2 + p_3) \times (p_1 + p_2 + p_3 + 1)/2$ vector obtained from $vec(\Sigma^\dagger)$ by eliminating all supradiagonal elements of the $(p_1 + p_2 + p_3) \times (p_1 + p_2 + p_3)$ matrix $\Sigma^\dagger$.

Denote by $f(\lambda; \theta)$ the spectral density matrix of the process $\{x(t), y(t), z(t)\}$ parametrized by the parameter θ. Assume that $\int_{-\pi}^{\pi} \log \det f(\lambda; \theta)d\lambda$ is differentiable with respect to θ and, for each λ, $f(\lambda; \theta)^{-1}$ is differentiable with respect to θ. The derivatives are denoted, respectively, by $H_j(\theta) = \partial \int_{-\pi}^{\pi} \log \det f(\lambda; \theta)d\lambda/\partial\theta_j$ and $h_j(\lambda; \theta) = \partial f^{-1}(\lambda; \theta)/\partial\theta_j$. Let $H(\theta)$ and $\text{tr}\{h(\lambda; \theta)f(\lambda)\}$ represent, respectively, the n_θ-vectors whose jth elements are $H_j(\theta)$ and $\text{tr}\{h_j(\lambda; \theta)f(\lambda)\}$. Let $S_{Tj}(\theta)$ be defined as

$$S_{Tj}(\theta) = H_j(\theta) + \int_{-\pi}^{\pi} \text{tr}\{h_j(\lambda; \theta)I_T(\lambda)\}d\lambda, \quad j = 1, \ldots, n_\theta,$$

where $I_T(\lambda)$ denotes the periodogram for $\{x(t), y(t), z(t), t = 1, ..., T\}$. Let $S_T(\theta)$ be the vector $vec(S_{Tj}(\theta), j = 1, ..., n_\theta)$. The value $\hat{\theta}$ satisfying the relation $S_T(\hat{\theta}) = 0$ is termed the maximum Whittle likelihood estimate of θ (see Whittle (1952, 1953) for scalar-valued and multiple stationary time series, respectively). Then, among other regularity conditions, we assume that θ is identifiable. Specifically, the identifiability condition of θ is formulated as: $R(\theta) \equiv H(\theta) + \int_{-\pi}^{\pi} \text{tr}\{h(\lambda; \theta)f(\lambda)\}d\lambda$ has a unique zero at $\theta = \theta_0$, where θ_0 is an interior point of the parameter space Θ, which is a compact subset of R^{n_θ}; see Assumption C(iii) of Hosoya (1997, p.117).

Conventional nonrestrictive estimation procedures for the VARMA model parameter do not necessarily produce estimates satisfying the zero conditions of det $A(z)$ and det $B(z)$. By modifying the maximum Whittle likelihood estimation, we suggest a three-step root modification procedure that produces coefficient estimates warranting stationarity and invertibility conditions (see Takimoto and Hosoya 2004, 2006 and Appendix A.2). The estimation is carried out by the following procedure:

Step A.1 By fitting a sufficiently higher-order VAR process and applying the ordinary least-squares method in the time domain, obtain an estimate of the unobservable disturbance terms as the regression residual series. For the case in which DGP is the VAR process, this step is skipped.

Step A.2 Substitute the disturbances in the MA part by the corresponding residuals obtained in Step A.1 and estimate the VARMA model using the time-domain least-squares method, selecting the lag order of the model with a certain information criterion. If the parameter values in Step A.2 do not satisfy the zero condition of $A(z)$, they are modified using the root contraction method in Appendix A.2.

Step A.3 Determine the estimate $\hat{\theta}$ of the model parameter by maximizing the Whittle likelihood endowed with a penalty function of the zero conditions. The maximizing algorithm is a quasi-Newton iteration method using the parameter values obtained in Step A.2 for the initial value of the Step A.3 iteration.

These first two steps without the explicit use of order choice criterion were originally proposed by Durbin (1960). The third step is a modified version of Takimoto and Hosoya (2004, 2006), and Sect. 4.1.2 provides details on the third-step algorithm. By setting the penalty asymptotically negligible, the conventional asymptotic normality holds for $\sqrt{T}(\hat{\theta} - \theta)$ under standard regularity conditions for the VARMA model (4.1) (see, e.g., Hannan and Rissanen (1982), Hannan and Kavalieris (1984a, b), and Hosoya (1997) for regularity conditions). Additionally, Dufour and Pelletier (2011) discussed the identifiability condition of the VARMA model. The modified maximum Whittle likelihood estimate for observation size T determined by these steps is denoted in the sequel by $\hat{\theta}$. Existing estimation methods for the VARMA model are characterized as follows:

- The estimation methods are classified into time-domain versus frequency-domain methods, and frequency-domain methods are able to exploit a longer sample-stretch (see Appendix A.3).
- Time-domain methods often use the repeated application of (modified) linear regressions, whereas the Whittle likelihood method needs to rely on an application of a certain nonlinear likelihood maximization algorithm. Estimations using linear regression generally produce stable coefficient estimates, whereas the mere application of a versatile optimization program does not necessarily provide stable estimates; iterations often produce estimates that drift away from the true value. It is possible that although an $(i+1)$th step estimate $\theta^{(i+1)}$ produces larger likelihood than the one in the ith step of the iteration, some zeros of $A^{(i+1)}(z)$ do not satisfy

the stationary condition. That is, there are some zeros inside the unit circle, where $A^{(i+1)}(z)$ is evaluated on the basis of $\theta^{(i+1)}$. One way to prevent the drift estimates is to attach a penalty term to the objective function, where the penalty is a function of the distance between the zeros of $A(z)$ and the unit circle, as proposed by Takimoto and Hosoya (2004, 2006).

- Typical regression approaches lack imposition of root condition requirements. It is noteworthy that all of the estimation procedures proposed by Hannan and Rissanen (1982), Hannan and Kavalieris (1984a, b), and Johansen (1995) have no built-in checking apparatus to prevent the root conditions from being violated.
- To impose an identifiability condition on the VARMA model, we need to introduce an artificial complication in the modeling.

Finally, there seems to remain the crucial problem of how to fill the gap (or to compromise) between small sample performance improvement and asymptotic–theoretic plausibility. For instance, take a standard circumstance, in which a third-step modified regression estimate such as Hannan's has the same (first-order) asymptotic distribution as the maximum-likelihood estimate. If the monitored likelihood value for a third-step modified regression estimate does not improve the corresponding value of the second-step (A.2) estimate of the three-step procedure, is the third-step estimate still used? Alternatively, is an estimate with greater likelihood preferred? There seems no easy solution to this problem.

4.1.2 Optimization Algorithm in Step 3

This section describes in detail the Step A.3 procedure. To simplify notations and for generality, this section considers the minimization of $f(x)$ with respect to x, where $f(x)$, for example, denotes the minus log Whittle likelihood function, indicated by x the n-vector comprising all model parameters.

Suppose that a second-order continuously differentiable function $f(x)$ is locally minimized for $x^* = \arg\min f(x)$. By expanding $f(x^*)$ around x, we have the approximation

$$0 = \left.\frac{\partial f(x)}{\partial x}\right|_{x=x^*} \approx \frac{\partial f(x)}{\partial x} + \frac{\partial^2 f(x)}{\partial x \partial x'}(x^* - x).$$

Hence, we have the approximation $x^{(\cdot)}$ for x^* in the neighborhood of x^*:

$$x^{(\cdot)} = x - \left[\frac{\partial^2 f(x)}{\partial x \partial x'}\right]^{-1} \frac{\partial f(x)}{\partial x}, \qquad (4.2)$$

where we set

$$\delta(x) = \left[\frac{\partial^2 f(x)}{\partial x \partial x'} \right]^{-1} \frac{\partial f(x)}{\partial x}.$$

Denote by $x^{(i)}$ and $x^{(i-1)}$ the ith and $(i-1)$th step approximations of x^* for $i = 1, 2, \cdots$. In view of (4.2), we have the recursive formula

$$x^{(i)} \equiv x^{(i)}(x^{(i-1)}, \lambda)$$

$$= x^{(i-1)} - \lambda \left[\frac{\partial^2 f(x)}{\partial x \partial x'} \bigg|_{x=x^{(i-1)}} \right]^{-1} \frac{\partial f(x)}{\partial x} \bigg|_{x=x^{(i-1)}}$$

$$\equiv x^{(i-1)} - \lambda \delta(x^{(i-1)}), \tag{4.3}$$

where the step-length parameter $\lambda \in (0, 1]$, given $x^{(i-1)}$, is determined to minimize $f(x^{(i)})$. In the sequel, the procedure to find such a scalar-valued λ is termed a linear search. Our procedure for minimizing the objective function $f(x)$ consists of an iterative application of the formula (4.3).

Let $x^{(0)}$ be an initial estimate of x. In the first step of the iteration, that is, ITER $= 1$, by setting the Hessian matrix equal to the identity matrix, the linear search is conducted to find $x^{(1)}$ using the steepest descent method, where the renovation formula is given by

$$x^{(1)} = x^{(0)} - \lambda \frac{\partial f(x)}{\partial x} \bigg|_{x=x^{(0)}}.$$

In the second step of the iteration, that is, ITER $= 2$, the Hessian matrix $H^{(1)}$ is evaluated for $x^{(1)}$ by the application of the Broyden–Fletcher–Goldfarb–Shanno (BFGS) formula, which is a Hessian matrix approximation formula, thus avoiding direct evaluation of a second derivative of $f(x)$ (e.g., see Judd 1999, pp.114–115). Through a linear search with $x^{(1)}$ and $H^{(1)}$, identify $x^{(2)}$. First, compare $f(x^{(2)})$ with $f(x^{(1)})$, where $x^{(1)}$ and $x^{(2)}$ are the steepest descent estimate and the quasi-Newton–Raphson estimate, respectively. When $f(x^{(1)}) < f(x^{(2)})$, we have two scenarios. One is that in the case of $f(x^{(0)}) < f(x^{(1)})$, the initial estimate $x^{(0)}$ is the final outcome in the minimization problem. Otherwise, the steepest descent estimate $x^{(1)}$ is our final outcome.

When we have an opposite inequality, such as $f(x^{(2)}) < f(x^{(1)})$, compare $f(x^{(2)})$ with $f(x^{(0)})$. If $f(x^{(0)}) < f(x^{(2)})$, the final outcome is $x^{(0)}$. Otherwise, check whether or not $x^{(2)}$ is converged. The final outcome is $x^{(2)}$ if convergence conditions are satisfied; otherwise, go to the third step of the iteration (ITER $= 3$). For the ith step of the iteration for $i = 3, 4, \cdots$, our Newton recursive algorithm consists of five substeps, which are given as follows:

Step B.1 Evaluate the Hessian matrix $H^{(i-1)}$ for $x^{(i-1)}$ by the application of the BFGS formula.

Step B.2 Find $x^{(i)}$ by a linear search using $x^{(i-1)}$ and $H^{(i-1)}$.

Step B.3 Compare $f(x^{(i)})$ with $f(x^{(i-1)})$. If $f(x^{(i-1)}) < f(x^{(i)})$, the final outcome is $x^{(i-1)}$ and go to Step B.5. Otherwise, go to Step B.4.

Step B.4 If the termination check is passed, the final outcome is $x^{(i)}$. Otherwise, set $i = i + 1$ and go to Step B.1.

Step B.5 Check the invertibility condition of the model on the basis of the final outcome. If it is not satisfied, create the invertible MA representation using the spectral factorization of Hosoya and Takimoto (2010).

These procedures are summarized in Fig. 4.1.

4.1.3 Monte Carlo Wald Test of Measures of Interdependence

Let $G(\theta)$ be an m-vector whose components are, respectively, certain distinct quantities related to the partial measures of interdependence. We classify tests on $G(\theta)$ into two classes according to the hypothesis type to be tested and propose different test statistics for respective classes. Because all of the interdependency measures are nonnegative quantities, a null measure constitutes an extreme value. To test the nullity of a set of measures, the direct use of the stochastic expansion of the estimates is not pertinent because the Jacobian matrix is not of full rank. Therefore, we need to separately address the following three cases:

Case 1: The test hypothesis does not involve the nullity of measures.
Case 2: Testing the nullity of measures.
Case 3: A mixed hypothesis of these two.

This book does not address Case 3. We provide a new test for the hypothesis of partial non-causality in Sect. 4.1.4, which is for Case 2. For another approach, see Sect. 5.1.2. First, we address Case 1. Suppose specifically that $G_i(\theta, \lambda), i = 1, \cdots, m$ are different types of scalar-valued measures, as exhibited in D.1 through D.4 in Sect. 3.3.2, and let $G(\theta, \lambda)$ be an m-vector such that $G(\theta, \lambda) = (G_1(\theta, \lambda), \ldots, G_m(\theta, \lambda))^*$, and $G_i(\theta, \lambda) > 0$. By stochastic expansion, we have

$$\sqrt{T}\{G(\hat{\theta}, \lambda) - G(\theta, \lambda)\} = (D_\theta G(\theta, \lambda))\sqrt{T}(\hat{\theta} - \theta) + o_p(1),$$

where $D_\theta G(\theta, \lambda)$ is the $m \times n_\theta$ Jacobian matrix of $G(\theta, \lambda)$ evaluated at θ; n_θ denotes the size of the vector θ. Suppose that $\sqrt{T}(\hat{\theta} - \theta)$ is asymptotically normally distributed with mean 0 and covariance matrix $\Psi(\theta)$ (see Hosoya 1997 for a set of milder conditions for the consistency and the asymptotic normality). Then, $\sqrt{T}\{G(\hat{\theta}, \lambda) - G(\theta, \lambda)\}$ is asymptotically normally distributed with mean 0 and the $m \times m$ asymptotic covariance matrix, which is given by

$$H(\theta, \lambda) = D_\theta G(\theta, \lambda)\Psi(\theta)D_\theta G(\theta, \lambda)^*.$$

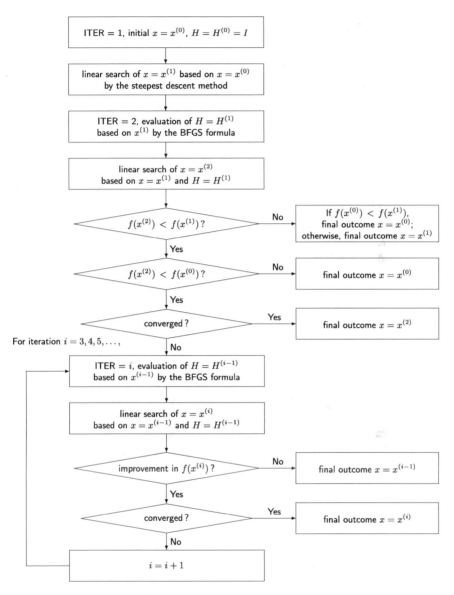

Fig. 4.1 Optimization in Step 3

Assume that the vector $G(\theta, \lambda)$ consisting of measures of interdependence is chosen such that rank $H(\theta, \lambda) = m$ in a neighborhood of the true θ. Then, the Wald statistic

$$W^{(m)}(\lambda) \equiv T\{G(\hat{\theta}, \lambda) - G(\theta, \lambda)\}^* H(\hat{\theta}, \lambda)^{-1}\{G(\hat{\theta}, \lambda) - G(\theta, \lambda)\}$$

is asymptotically χ^2-distributed with m degrees of freedom if θ is the true value. Let G_0 be a given m vector. Then, the null hypothesis $G(\theta, \lambda) = G_0$ can be tested using the test statistic

$$W^{(m)}(\lambda) \equiv T\{G(\hat{\theta}, \lambda) - G_0\}^* H(\hat{\theta}, \lambda)^{-1}\{G(\hat{\theta}, \lambda) - G_0\},$$

where G_0 is a vector of positive components. Additionally, a confidence set for $G(\theta, \lambda)$ can be constructed using the statistic $W^{(m)}(\lambda)$.

There are several alternative procedures available to estimate the asymptotic covariance matrix $H(\theta, \lambda)$. For example, we might use the asymptotic covariance matrix evaluation formula given by Yao and Hosoya (2000) for the cointegrated VAR model; however, the formula becomes much more involved computationally for the general ARMA model.

Alternatively, a simpler approach is to use the Monte Carlo Wald test procedure which is conducted as follows. For simplicity, we assume that $\theta = vec\{A, B, v(\Sigma^\dagger)\}$.

Step C.1 Estimate θ using the modified maximum Whittle likelihood method and obtain the vector $G(\hat{\theta}, \lambda)$.
Step C.2 Generate the data series $\{x(t)^\dagger, y(t)^\dagger, z(t)^\dagger; t = 1, \cdots, T\}$ using the model (4.1) in which the model parameter is substituted by the estimate $\hat{\theta}$ obtained in Step C.1, and the disturbance terms are simulated as independently and normally distributed random vectors $\{\varepsilon(t)\}$ with mean 0 and the estimated covariance matrix $\hat{\Sigma}^\dagger$ in Step C.1.
Step C.3 Estimate the parameter θ using the simulated data series $\{x(t)^\dagger, y(t)^\dagger, z(t)^\dagger; t = 1, \cdots, T\}$ and set the estimate of $G(\theta, \lambda)$ by $G(\theta^\dagger, \lambda)$.
Step C.4 Iterate Steps C.2 and C.3 N times, produce $G(\theta_n^\dagger, \lambda); n = 1, \cdots, N$, and estimate the covariance matrix $H(\theta, \lambda)$, denoting the estimate as $\hat{H}(\hat{\theta}, \lambda)$, as the Monte Carlo sample covariance matrix of $G(\theta_n^\dagger, \lambda)$. Namely,

$$\hat{H}(\hat{\theta}, \lambda) = \frac{T}{N} \sum_{n=1}^{N} \left(G(\theta_n^\dagger, \lambda) - \bar{G}(\theta^\dagger, \lambda)\right) \left(G(\theta_n^\dagger, \lambda) - \bar{G}(\theta^\dagger, \lambda)\right)^*,$$

where $\bar{G}(\theta^\dagger, \lambda) = N^{-1} \sum_{n=1}^{N} G(\theta_n^\dagger, \lambda)$.

Remark 4.1 It is difficult to provide a general rule to relate the simulation size of N to the observation size T. However, in practice it is not difficult to determine an appropriate size N by inspecting how the calculated covariance matrices are numerically stabilized as the number N increases through Monte Carlo simulation.

4.1.4 Monte Carlo Wald Testing of Non-causality

As previously explained, the foregoing approach is not suited for testing non-causality; therefore, we must look for other statistics. To test non-causality, Breitung and Candelon (2006) used bivariate stationary models as well as cointegrated VAR models to propose an F-test for a set of linear restriction hypotheses on certain distributed-lag parameters in their autoregressive distributed-lag (ARDL) model. However, to address a broader class, such as the VARMA model, a more general approach is required. The Breitung and Candelon test uses the standard asymptotic theory of the stationary time-series regression estimation and testing, whereas our Monte Carlo Wald test approach uses the standard asymptotic theory of the Whittle likelihood inference for stationary multivariate ARMA processes. Instead of forwarding explicitly a set of assumptions, we assume in the following arguments that the consistency and asymptotic normality of the subject-matter statistics $G(\hat{\theta}, \lambda)$ used in the Wald tests hold and that the Monte Carlo estimate of the covariance matrix of the statistics based on Monte Carlo iterations is consistent. Although the Gaussian pseudorandom number series is the most convenient choice to simulate observation series, for a more sophisticated approach, we may as well apply certain time-series bootstrap methods.

Because the formula (3.10) implies that the measure $PM_{y \to x:z}(\lambda)$ is not determined by $\Gamma_{12}^{\dagger}(e^{-i\lambda})$ alone, but is determined in terms of the ratio $\Gamma_{11}^{\dagger}(e^{-i\lambda})^{-1} \Gamma_{12}^{\dagger}(e^{-i\lambda})$, we may conduct the test of the null hypothesis of $\{v(t)\}$ not causing $\{u(t)\}$ at frequency λ by testing

$$\Gamma_{11}^{\dagger}(e^{-i\lambda})^{-1} \Gamma_{12}^{\dagger}(e^{-i\lambda}) = 0, \tag{4.4}$$

instead of testing the hypothesis $\Gamma_{12}^{\dagger}(e^{-i\lambda}) = 0$, where $\Gamma_{kl}^{\dagger}(e^{-i\lambda}) \equiv \sum_{j=0}^{\bar{a}} \Gamma_{kl}^{\dagger}[j] e^{-ij\lambda}$ and $\Gamma_{kl}^{\dagger}[j]$ is the jth coefficient matrix of the polynomial $\Gamma_{kl}^{\dagger}(z)$, for $k, l = 1, 2$ and $\bar{a} \equiv a(p_1 + p_2 + p_3 - 1) + b$. Define

$$\psi(\theta, \lambda) \equiv vec\{Re\, \Gamma_{11}^{\dagger}(e^{-i\lambda})^{-1} \Gamma_{12}^{\dagger}(e^{-i\lambda}), Im\, \Gamma_{11}^{\dagger}(e^{-i\lambda})^{-1} \Gamma_{12}^{\dagger}(e^{-i\lambda})\},$$

where $Re\,A$ and $Im\,A$ indicate the real and imaginary parts of the complex-valued matrix A, respectively. Then, the Wald statistic for the null hypothesis that $y(t)$ does not cause $x(t)$ at frequency λ in the presence of $z(t)$ is given by

$$W^{(n)}(\lambda) = T(\psi(\hat{\theta}, \lambda))^* H(\hat{\theta}, \lambda)^{-1} \psi(\hat{\theta}, \lambda), \tag{4.5}$$

where $H(\theta, \lambda)$ is the asymptotic covariance matrix of $\sqrt{T}(\psi(\hat{\theta}, \lambda) - \psi(\theta, \lambda))$. It is certainly possible to construct the test statistics for the null hypothesis of $\Gamma_{12}^{\dagger}(e^{-i\lambda}) = 0$ alone, and in the next section, we compare the test performance of two approaches using two trivariate series. For the evaluation of $H(\theta, \lambda)$, one approach is to apply the Case 1 method using stochastic expansion. The approach is useful in case the covariance matrix of $\psi(\hat{\theta}, \lambda)$ is numerically tractable. Another approach to evaluating the covariance matrix $H(\theta, \lambda)$ without relying on stochastic expansion is to apply the Monte Carlo procedure Steps C.1 through C.4 given in Sect. 4.1.3 directly.

To address testing the null hypothesis of the overall measure of the one-way effect $M_{v \to u} = 0$, namely $\{v(t)\}$ not causing $\{u(t)\}$, there are several approaches to the test. The component $\Gamma_{12}^{\dagger}(z)$ in (3.9) has a finite-order MA expression such that $\Gamma_{12}^{\dagger}(z) = \sum_{j=0}^{\bar{a}} \Gamma_{12}^{\dagger}[j, \theta] z^j$. Then, testing $vec(\Gamma_{12}^{\dagger}[j, \theta], j = 0, \cdots, \bar{a}) = 0$ does not constitute a boundary value test. In the case of $p_1 > 1$, another method to test the null OMO is to test $\Gamma_{11}^{\dagger}(z)^{-1} \Gamma_{12}^{\dagger}(z) = 0$. The test is reduced to the test of

$$vec(\Xi[j, \theta], j = 0, \cdots, \bar{a} p_1) = 0,$$

where $\Xi[j, \theta]$ is determined by the equality

$$\Gamma_{11}^{\dagger}(z, \theta)^{\sharp} \Gamma_{12}(z, \theta) = \sum_{j=0}^{\bar{a} p_1} \Xi[j, \theta] z^j,$$

where $\Gamma_{11}^{\dagger}(z, \theta)^{\sharp}$ is the $p_1 \times p_1$ upper left submatrix of the $(p_1 + p_2) \times (p_1 + p_2)$ adjugate matrix $\Gamma^{\dagger}(z, \theta)^{\sharp}$. For those two tests, we can apply the Wald test approach using the modified Whittle estimator $\hat{\theta}$ and the relevant covariance matrix estimate. A third candidate is to test

$$\frac{1}{2\pi} \int_{-\pi}^{\pi} PM_{y \to x:z}(\lambda) d\lambda = 0,$$

where the integrand is defined in (3.10). Although we can numerically evaluate the integral for the estimated $\hat{\theta}$, the test constitutes a boundary value test and standard large-sample test techniques do not apply, as remarked in Sect. 4.1.3. To test the nullity of the overall measures in Sects. 4.2 and 4.3, we construct the Wald statistics for testing $\Gamma_{12}^{\dagger}(z) = 0$ because, for numerical illustrations, we focus on the case of $p_1 = 1$ in this chapter.

Remark 4.2 Suppose that $u(t)$ and $v(t)$ are scalar-valued and are generated by the bivariate AR process:

$$u(t) = \sum_{j=1}^{a} \alpha_1[j] u(t - j) + \sum_{k=1}^{a} \beta_1[k] v(t - k) + \varepsilon_1(t),$$

$$v(t) = \sum_{j=1}^{a} \alpha_2[j] u(t - j) + \sum_{k=1}^{a} \beta_2[k] v(t - k) + \varepsilon_2(t).$$

For such a model, Breitung and Candelon (2006) noted that the test of $\Gamma_{12}^{\dagger}(e^{-i\lambda}) = 0$ in (3.10) to test $\{v(t)\}$ not simply causing $\{u(t)\}$ is equivalent to testing

$$\Gamma_{12}^{\dagger}(e^{-i\lambda})^{\sharp} = 0, \tag{4.6}$$

where $\Gamma_{12}^{\dagger}(e^{-i\lambda})^{\sharp}$ is the (1,2) component of the 2×2 adjugate matrix $\Gamma^{\dagger}(e^{-i\lambda})^{\sharp}$; however, the test (4.6) is reduced to testing

$$\sum_{k=1}^{a} \beta_1[k]e^{-ik\lambda} = 0. \tag{4.7}$$

Testing the hypothesis (4.7) can be addressed using an F-test because it imposes linear restrictions on the distributed-lag coefficients. However, this method does not extend to a higher-dimensional $\Gamma^{\dagger}(e^{-i\lambda})$ because (4.6) imposes nonlinear restrictions on the model parameters. Accordingly, a certain version of either the likelihood ratio test or the Wald test of nonlinear restrictions, rather than the F-test, is required for the more general case.

Remark 4.3 Breitung and Candelon (2006, p.369) proposed a way to eliminate a third series effect using a time-domain regression. They proposed fitting a single-equation ARDL model

$$x(t) = \sum_{j=1}^{a} \alpha[j]x(t-j) + \sum_{k=1}^{a} \beta[k]y(t-k) + \sum_{l=1}^{a} \gamma[l]v(t-l) + \varepsilon(t) \tag{4.8}$$

and to test the null hypothesis $\sum_{k=1}^{a} \beta[k]e^{-ik\lambda} = 0$ using an F-statistic, where $v(t)$ is equal to either $z(t)$ or the residual obtained by regressing $z(t)$ on $x(t)$, $y(t)$, and $w(t-1), \cdots, w(t-a)$, where $w = (x, y, z)^*$. They presented an empirical analysis of the one-way effect of the yield spread on the real GDP growth rate in the USA by eliminating the real balance effect in the time domain and concluded that the test results do not depend on the choice of $v(t)$ in (4.8).

4.2 Simulation Performance

4.2.1 Designing Monte Carlo Simulation

This section investigates the small sample performance of the partial one-way effect measure estimates based on Monte Carlo data generated by a pair of trivariate VAR(2) and VARMA(1,1) models. To estimate the partial frequency-wise measures of interdependence, spectral factorization is conducted for the spectral density $h(\lambda)$ given in (3.6), and the parameters are substituted by the modified Whittle estimate presented in Sect. 4.1.1. Using an eigenvalue contraction method accompanied by the canonical factorization of the MA spectrum, our procedure produces, in each of Steps A.2 and A.3, VARMA coefficient estimates for which the stationarity and invertibility conditions are satisfied. For the Monte Carlo replications, we set the number of replications to 1000 times, and in each of which the sample size T is set to either 100 or 1500. The obtained Monte Carlo sample produced by 1000 replications is used to evaluate the mean, the standard error, and the covariance matrix. To eliminate the initial-value influence, we discard the first 100 values of the generated vector values.

Given a three variable system $\{x(t), y(t), z(t)\}$, we use the following notations henceforth:

a. FMO and FMO2 denote the partial frequency-wise measures of the one-way effect given $z(t)$ as the third series; to be explicit, FMO $\equiv$ FMO$(y \rightarrow x|z)$, FMO2 $\equiv$ FMO$(x \rightarrow y|z)$. Similarly, OMO and OMO2 denote the corresponding partial overall measures: OMO $\equiv$ OMO$(y \rightarrow x|z)$, OMO2 $\equiv$ OMO$(x \rightarrow y|z)$.

b. FW and FW2 denote the frequency-wise Wald statistics for testing y not causing x (x not causing y, respectively) with z as the third series. Namely, FW $\equiv$ FW$(y \nrightarrow x|z)$ and FW2 $\equiv$ FW$(x \nrightarrow y|z)$. Similarly, the overall Wald statistics are denoted by OW $\equiv$ OW$(y \nrightarrow x|z)$ and OW2 $\equiv$ OW$(x \nrightarrow y|z)$.

c. The same notation is used in Sect. 4.3.2. For example, in case $x \equiv \triangle GDP$, $y = TS$, $z = (\triangle M2, \triangle CPI)$, we set FW2 $\equiv$ FW$(\triangle GDP \nrightarrow TS|\triangle M2, \triangle CPI)$.

In addition to these measures, FMR and FMA are also shown in the figures, where FMR and FMA indicate the frequency-wise partial measures of reciprocity and association, respectively. Additionally, OMR and OMA in the tables denote the corresponding partial overall measures.

Model I: Suppose that a stationary trivariate VAR(2) model is given by

$$
\begin{bmatrix} x(t) \\ y(t) \\ z(t) \end{bmatrix} = A[1] \begin{bmatrix} x(t-1) \\ y(t-1) \\ z(t-1) \end{bmatrix} + A[2] \begin{bmatrix} x(t-2) \\ y(t-2) \\ z(t-2) \end{bmatrix} + \varepsilon(t), \tag{4.9}
$$

where $\{\varepsilon(t)\}$ is an independently, identically, and normally distributed sequence. Namely $\varepsilon(t) \sim i.i.d.\ N(0, \Sigma_\varepsilon)$, and

$$
A[1] = \begin{bmatrix} 0.905 & -0.023 & 0.450 \\ 0.234 & 0.714 & 0.360 \\ 0.147 & -0.053 & 1.171 \end{bmatrix}, \qquad A[2] = \begin{bmatrix} -0.15 & 0.20 & -0.45 \\ -0.19 & 0.08 & -0.30 \\ -0.11 & 0.05 & -0.32 \end{bmatrix},
$$

$$
\Sigma_\varepsilon = \begin{bmatrix} 0.47 & 0.20 & 0.18 \\ 0.20 & 0.32 & 0.27 \\ 0.18 & 0.27 & 0.30 \end{bmatrix}.
$$

The coefficients and the covariance matrix are the same as the ones provided by Yap and Reinsel (1995, p.262) in their Case 1(b). The zeros of $\det(I - A[1]z - A[2]z^2)$ are equal to $-0.405 \pm 7.764i$, $1.266 \pm 0.039i$, 1.320, 2.135, whence the process (4.9) is second-order stationary.

Four true frequency-wise partial measures of interdependence—FMO, FMO2, FMR, and FMA—are evaluated using Model I and all of them are shown in Fig. 4.2a. For $\lambda \leq 0.38\pi$, FMO is larger than FMO2; otherwise, FMO2 is larger than FMO. Furthermore, FMA is relatively high on a low-frequency domain. Four true overall partial measures—OMO, OMO2, OMR, and OMA—are reported in Table 4.1. OMO

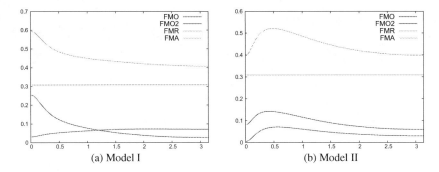

Fig. 4.2 True measures

is larger than OMO2, which is consistent with Fig. 4.2a. Figure 4.2a indicates that the contribution of the reciprocity measure in the association is notable.

Model II: The second data generating process we examine is the trivariate ARMA(1,1) model given as follows:

$$\begin{bmatrix} x(t) \\ y(t) \\ z(t) \end{bmatrix} = A[1] \begin{bmatrix} x(t-1) \\ y(t-1) \\ z(t-1) \end{bmatrix} + \varepsilon(t) + B[1]\varepsilon(t-1), \tag{4.10}$$

where

$$A[1] = \begin{bmatrix} 0.755 & 0.177 & 8.55 \times 10^{-5} \\ 0.044 & 0.794 & 0.060 \\ 0.037 & -0.003 & 0.851 \end{bmatrix}, \quad B[1] = \begin{bmatrix} -0.085 & -0.336 & -0.263 \\ -0.243 & -0.132 & -0.064 \\ -0.099 & 0.034 & -0.378 \end{bmatrix},$$

and the covariance matrix is the same as the one in Model I. The zeros of $\det(I_3 - A[1]z) = 0$ and $\det(I_3 - B[1]z) = 0$ are 1.111, 1.25, 1.429, and -4.951, 2, 3.367, respectively; therefore, the model generates a stationary invertible process. The MA coefficient matrix $B[1]$ is obtained by setting $\lambda_\theta = 0.5$ in Yap and Reinsel (1995, p.263).

The true measures for the model (4.10) are displayed in Fig. 4.2b. In contrast to the VAR(2) case, FMO dominates FMO2 on the entire frequency domain. FMA, FMO, and FMO2 have more or less similar tendencies. That is, for $\lambda < 0.13\pi$, they are increasing monotonically.

4.2.2   Simulation Results

Although we are not able to know the extent to which $\{y(t)\}$ causes $\{x(t)\}$ with respect to the prediction improvement using only the Granger causality test, the overall and

Table 4.1 Overall measures of the partial one-way effect

Model I	OMO	OMO2	OMR	OMA
True	0.074	0.064	0.309	0.447
Sample size = 100				
Mean	0.097	0.086	0.321	0.504
Mean−s	0.035	0.033	0.213	0.388
Mean+s	0.159	0.138	0.430	0.619
Sample size = 1500				
Mean	0.075	0.066	0.309	0.451
Mean−s	0.061	0.053	0.282	0.423
Mean+s	0.090	0.079	0.337	0.479
Model II	OMO	OMO2	OMR	OMA
True	0.093	0.045	0.309	0.447
Sample size = 100				
Mean	0.108	0.067	0.325	0.499
Mean−s	0.039	0.012	0.206	0.362
Mean+s	0.177	0.121	0.444	0.637
Sample size = 1500				
Mean	0.092	0.045	0.310	0.447
Mean−s	0.075	0.034	0.282	0.418
Mean+s	0.108	0.056	0.337	0.475

Note True denotes the true value of the measure, mean is the average of 1000
Monte Carlo replications, and s denotes the square root of the sample variance

frequency-wise measures proposed in this book provide a quantitative expression of
the one-way causal effect. To be more complete, the construction and presentation
of confidence sets of those measures are desired. This book does not address the
confidence set construction (for such an approach, see Yao and Hosoya (2000)). The
numerical results in this section are limited to p-values on the basis of overall and
frequency-wise Wald tests and estimates of the measures for the two models.

(1) Estimation performance of overall measures and test results

Table 4.1 indicates that the numerical performance of the OMO and OMO2 estimates
for the VAR(2) and VARMA(1,1) models follows more or less what we expect
from the asymptotic theory. The estimation precision in terms of bias and standard
deviation improves as the sample size increases.

Table 4.2 illustrates the four cases of Monte Carlo p-values based on OW and OW2
obtained for each of the combination of the two hypotheses and the two sample sizes
100 and 1500. Note that each Wald statistic value is the Monte Carlo test result
for one sample in testing $\Gamma_{12}^{\dagger}(z) = 0$ for $\Gamma_{12}^{\dagger}(z)$ given in (3.9). For the sample size
of 100, except for OW2 in Model II, tests do not lead to the rejection of the null
hypotheses at the 5% significance level, whereas for the sample size of 1500, all four

Table 4.2 Wald statistics for non-causality and the p-values

Model I	Statistic	DF	p-value
	Sample size=100		
OW	8.719	4	0.069
OW2	5.079	4	0.279
	Sample size=1500		
OW	114.047	4	0.000
OW2	103.102	4	0.000
Model II	Statistic	DF	p-value
	Sample size=100		
OW	3.695	3	0.296
OW2	10.444	3	0.015
	Sample size=1500		
OW	113.977	3	0.000
OW2	50.162	3	0.000

cases definitely reject the null hypothesis of non-causality. The table shows that the sample size of 100 is not enough to reject the null hypotheses even if they are wrong.

(2) Estimation performance of frequency-wise measures

For Model I, Fig. 4.3a–d plot the FMO and FMO2 for two sample sizes of 100 and 1500. Figure 4.3a and b show that FMO takes larger values on a low-frequency domain and becomes smaller as the frequency increases. The result is consistent with the tendency of the true FMO, which decreases monotonically as the frequency increases. For the sample size of 100, the mean values of FMO and FMO2 tend to overestimate the true values on the low-frequency domain. For FMO, the sample means of the estimates closely follow the true values on the entire domain. Even in the small sample setting, $T = 100$, the mean values of estimated FMOs are close to the true FMO. Regarding FMO2, although for the large-sample setting, $T = 1500$, the mean of FMO2 closely follows the true FMO2 in Fig. 4.3d, for the small sample size, Fig. 4.3c shows that the bias and the standard deviation increase monotonically as the frequency decreases. However, the true values of both measures are in the one standard deviation interval of the means for all frequencies. For both measures, as the large-sample theory predicts, the accuracy in terms of bias and standard deviation improves as the size increases from 100 to 1500.

The estimate performances for Model II as exhibited in Fig. 4.4b and d are consistent with the asymptotic theory, paralleling Model I. The mean values of FMO and FMO2 with the sample size of 1500 successfully reproduce the single peak feature.

The common feature for FMO and FMO2 and the two models is the increasing sample variation of the estimates as $\lambda \to 0$, especially in the small sample setting. Namely, the estimates become variable as the frequency decreases. Therefore, we

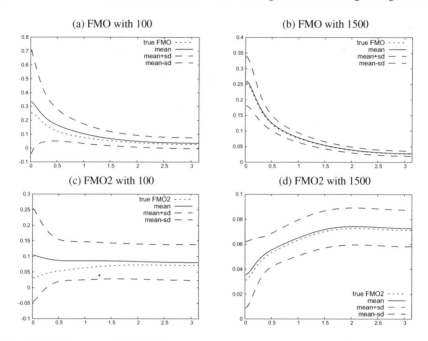

Fig. 4.3 Partial measures for Model I

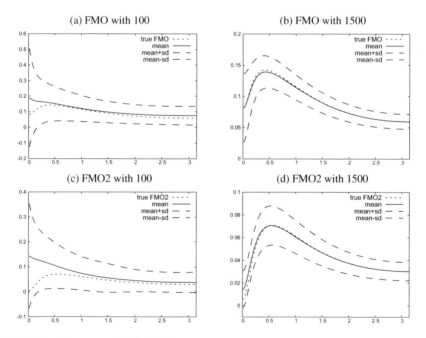

Fig. 4.4 Partial measures for Model II

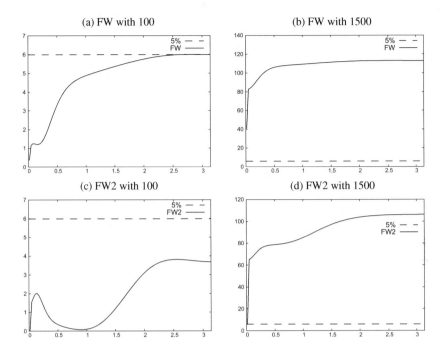

Fig. 4.5 Wald statistics for Model I

must take into account the sample variation of the one-way effect measure estimates on the low-frequency domain in empirical analyses.

(3) Performance of frequency-wise Wald tests

Figures 4.5 and 4.6 show the simulation results of the Wald statistics evaluated by formula (4.5) for Models I and II, respectively, where the null hypothesis is posited by (4.4). For each model and for each sample size, a one time-series sample is generated and the frequency-wise Monte Carlo Wald statistics are computed for that sample.

For Model I with the sample size of 100, the test statistic FW reaches the 5% significance level on the frequency band $\lambda > 0.83\pi$. Although the p-value for OW is 0.069, as given in Table 4.2, there exists the significant frequency domain of $\lambda > 0.83\pi$. The interpretation of periodicity depends on the data set. For example, for yearly data, $\lambda = 0.83\pi$ implies a 2.4-year period in terms of periodicity. FW2 does not reach the 5% significance level at all on the entire domain, which is consistent with the OW2 test result. For the large-sample setting, $T = 1500$, FW and FW2 are significant on the entire domain, which is consistent with the overall results given in Table 4.2.

Regarding the test performance for Model II, FW with $T = 100$ is well below the 5% significance level on the entire domain, whereas both FW and FW2 with $T = 1500$ as well as FW2 with $T = 100$ exceed the 5% level, except for very narrow bands, which are located in the vicinity of the origin. Figure 4.6 is consistent with

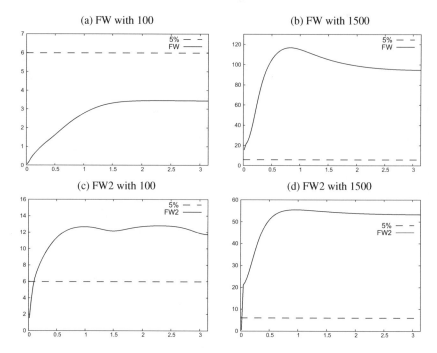

Fig. 4.6 Wald statistics for Model II

the result in Table 4.2 because only OW with $T = 100$ accepts the null hypothesis of no one-way effect at the 5% significance level; OW with $T = 1500$ and OW2 with $T = 100$ and 1500 reject the null hypothesis at the 5% significance level. For FW2 with $T = 100$, the Wald test result shows that, for yearly data, except for the cycle with a periodicity longer than 16.5 years, the frequency-wise partial measures are statistically significant.

The test results of $\Gamma_{12}^{\dagger}(e^{-i\lambda}) = 0$ are compared with the results of the test for $\Gamma_{11}^{\dagger}(e^{-i\lambda})^{-1}\Gamma_{12}^{\dagger}(e^{-i\lambda}) = 0$. Figures 4.7 and 4.8 which plot frequency-wise Wald statistics for the test of $\Gamma_{12}^{\dagger}(e^{-i\lambda}) = 0$ present very similar characteristics as the ones in Figs. 4.5 and 4.6, and we do not observe any notable difference between two Wald statistics, except for FW with $T = 100$ in Model I.

4.2.3 Comparison of Step 2 and Step 3 Estimation

This section compares Step 3 estimates with Step 2 estimates in the setting of Sect. 4.2.1 to examine whether our three-step estimation procedure works effectively. For that purpose, we pose the following questions:

(1) Is the root modification required in Step 2?
(2) Does the penalty term in the objective function properly work?

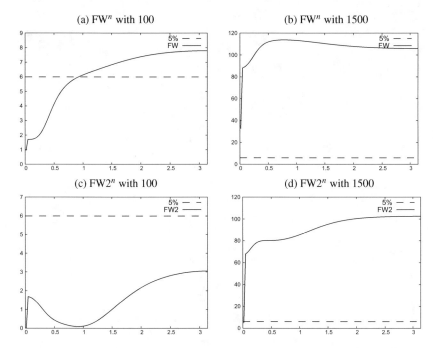

Fig. 4.7 Wald statistics based on the numerator alone for Model I. *Note* FW^n and FW2^n denote that the Wald statistics are constructed of the numerator alone to test non-causality in the frequency domain

(3) Is the likelihood improved in Step 3?

(4) How do the measure estimates in Step 3 compare with the ones in Step 2?

(1) Characteristics of Step 2 parameter estimation

When the initial values do not satisfy the stationarity condition, we cannot expect that all zeros of $A(z)$ based on the final outcome in Step 3 are outside the unit circle. In the normal setting of the nonlinear optimization algorithm, there is no automatic built-in algorithm to make the estimate satisfy the zero condition. In fact, the Step 2 estimates of VAR(2) and VARMA(1,1) with the sample size of 100 fail to satisfy the zero condition of $A(z)$ 2 and 92 times out of 1000 repetitions, respectively, even though the DGPs are stable and invertible. In such cases, it would be preferable to start from a new stable estimate value. We apply the root contraction method for that purpose (see Appendix A.2).

(2) Penalty term in the objective function of Step 3

When a stable initial value always produces a stable final outcome, we do not need to take into account the penalty term. Unfortunately, a desirable property of the starting value is not necessarily inherited in subsequent values. Regarding outcomes that fall in the region in which some zeros of $A(z)$ are inside the unit circle, the number of such

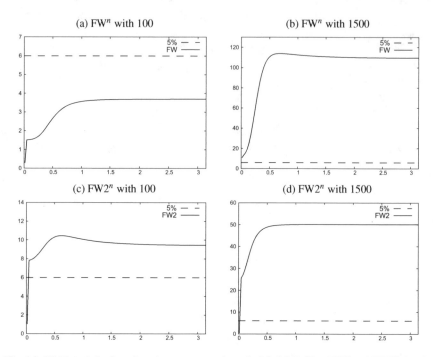

Fig. 4.8 Wald statistics based on the numerator alone for Model II. *Note* FW^n and $FW2^n$ denote that the Wald statistics are constructed of the numerator alone to test non-causality in the frequency domain

nonstable outcomes for VAR(2) with $T = 100$ and 1500 and for VARMA(1,1) with $T = 100$ and 1500 are 390, 452, 69, and 151 out of 1000 repetitions, respectively. By outfitting the objective function with the penalty term, we can avoid nonstable outcomes. Regarding the MA coefficient estimation, for VARMA(1,1) with $T = 100$, it is observed that 11 cases out of 1000 repetitions violate the invertibility condition. We can recover the invertible representation of the MA part by employing the spectral factorization procedure.

(3) Improvement of the objective function in Step 3

The maximization problem of the Whittle likelihood is reduced to minimizing $\log \det \hat{\Sigma}^\dagger$ (see Appendix A.3). We evaluate the average of the difference between $\log \det \hat{\Sigma}^\dagger_{Step2}$ and $\log \det \hat{\Sigma}^\dagger_{Step3}$. For VARMA(1,1) with the sample size of 100, the average improvement of the objective function in 1000 repetitions is 4.92×10^{-4}, which is the maximum value among four settings, whereas VAR(2) with the sample size of 1500 attains a minimum improvement of 9.98×10^{-9}. For VAR(2) with $T = 100$ and VARMA(1,1) with $T = 1500$, the differences between two $\log \det \hat{\Sigma}^\dagger$ are 9.80×10^{-6}, and 1.64×10^{-6}, respectively. In view of these results, Step 3 of our three-step procedure seems to work effectively to attain a smaller value of the objective function.

(4) Comparison of FMO and FMO2 estimates in Steps 2 and 3

The estimates of FMO and FMO2 are evaluated for $\lambda_j = (j/B)\pi$ and $j = 1, 2, \cdots, B$. In the comparison, set $B = 200$ and $N = 1000$ in the sequel. The estimates of measures in the ith repetition and the true values of FMO or FMO2 are denoted by $\hat{M}_i$ and M^0, respectively. The criteria for deviations in each frequency are denoted by the mean absolute error (MAE) and the root-mean-squared error (RMSE):

a. $\text{MAE} \equiv \dfrac{1}{N} \sum_{i=1}^{N} |\hat{M}_i(\lambda) - M^0(\lambda)|$

b. $\text{RMSE} \equiv \sqrt{\dfrac{1}{N} \sum_{i=1}^{N} (\hat{M}_i(\lambda) - M^0(\lambda))^2}$

To examine the average deviation on the entire frequency domain, the average MAE (AMAE) and the average RMSE (ARMSE) are calculated as criteria:

c. $\text{AMAE} \equiv \dfrac{1}{B} \sum_{j=1}^{B} \dfrac{1}{N} \sum_{i=1}^{N} |\hat{M}_i(\lambda_j) - M^0(\lambda_j)|$

d. $\text{ARMSE} \equiv \dfrac{1}{B} \sum_{j=1}^{B} \sqrt{\dfrac{1}{N} \sum_{i=1}^{N} (\hat{M}_i(\lambda_j) - M^0(\lambda_j))^2}$

We evaluate the extent to which Step 3 of our three-step procedure affects estimates of the one-way effect measure based on the difference in those criteria.

Table 4.3 reports the averages of MAE(RMSE) for Step 2 minus MAE(RMSE) for Step 3. Using the average criteria, we can observe that for VAR(2) the optimization in Step 3 improves estimates. In seven out of eight cases, measure estimates are

Table 4.3 Overall differences in Steps 2 and 3 estimates

Criterion	Sample	FMO	FMO2
		Model I	
AMAE	100	3.61^{-4}	6.66^{-4}
AMAE	1500	3.67^{-8}	6.25^{-7}
ARMSE	100	-3.14^{-5}	1.54^{-4}
ARMSE	1500	5.33^{-8}	2.11^{-8}
		Model II	
AMAE	100	-2.48^{-3}	2.59^{-4}
AMAE	1500	1.17^{-4}	9.15^{-5}
ARMSE	100	-1.45^{-3}	-2.96^{-4}
ARMSE	1500	7.57^{-6}	3.43^{-6}

Note AMAE and ARMSE denote the average mean absolute error and the average root-mean-squared error, respectively

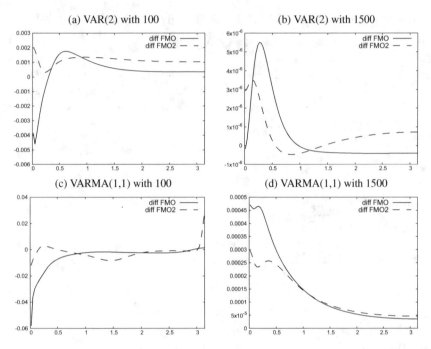

Fig. 4.9 Frequency-wise differences of RMSE in Steps 2 and 3 estimates. *Note* The diff FMO and diff FMO2 denote that $\hat{M}_i(\lambda)$ is set as FMO and FMO2, respectively

closer to the true values in Step 3 rather than in Step 2 because positive AMAE and ARMSE indicate that the average deviation from the true values is smaller in Step 3. For VARMA(1,1) with the sample size of 100, in three out of four cases, measure estimates in Step 2 are closer to the true values. However, with the sample size of 1500, Step 3 estimates in all four cases are closer to the true values. For the small sample cases, the Step 3 procedure does not necessarily improve the estimation, although likelihoods are improved.

Figure 4.9 shows differences in the Step 2 and 3 RMSE for each frequency. For example, in Fig. 4.9a, which deals with VAR(2) with the sample size of 100, we observe that except for the case of FMO2 for $\lambda < 0.1\pi$, Step 3 estimates of the one-way effect are closer than the Step 2 estimates to the true values. However, the deviation in the Step 3 estimates becomes larger for $\lambda < 0.1\pi$. Moreover, because the impact for $\lambda < 0.1\pi$ is relatively larger, ARMSE for FMO with $T = 100$ takes negative values in Table 4.3. For the sample size of 1500, Fig. 4.9b shows that both of Step 2 and Step 3 estimates closely approximate the true values. We can conclude that the Step 3 iteration produces more accurate estimates of the one-way effect measures for the VAR(2) model.

Figure 4.9c and d indicate that the findings for VAR(2) model hold for VARMA(1,1) model. For VARMA(1,1) with $T = 1500$, all FMO and FMO2 estimates in Step 3 are closer to the true values, and the results are consistent with the asymptotic theory's prediction. In view that the scales of the vertical axes in Figs. 4.9c and d are larger than those in Figs. 4.9a and b, respectively, the benefit from employing the Step 3 iteration seems higher by the involvement of the MA part in the DGP.

To summarize, in view of the frequency-wise criteria, some of Step 2 measures estimates outperform those in Step 3 for the small sample, especially on the lower frequency domain, whereas for the large sample, Step 3 optimization works well but also it does not improve Step 2 for some frequencies.

4.3 Empirical Analysis of Macroeconomic Series

4.3.1 Literature

4.3.1.1 Macroeconomic Causal Relations

Sims (1980) conducted his "innovation accounting" analysis on the basis of the reduced moving average form whose innovation terms have diagonal covariance matrix obtained through triangular matrix multiplication. He investigated two US data sets on the interwar period 1920–41 and the postwar period 1948–78. Each data set consists of monthly observations on four variables: industrial production (y), wholesale price index (p), money stock M1(m), and short-term interest rate (r); all variables are log-transformed data. In his framework, ordering the components of the vector (y, p, m, r) is crucial and must be determined in view of the causal direction in question between two series in the presence of confounding series. Sims investigates the causal relations on the basis of the MA representation (3.11) and its estimation and allied significance testing. For instance, Table 1 of Sims (1980, p.252) employs the ordering (x, y, z) $=$ (m, y, p) for the three-variate model, whereas Tables 2 and 3 are based on the ordering for the four-variate model. Monetarists have the view that monetary policy has central importance in the business cycle, where monetary policy is represented by the time path of money stock, asserting that the pure innovation component of money stock causes a change in output. In contrast, non-monetarists direct their attention to the role of future expectations in investment behavior, specifically the future profitability of capital to be purchased. According to Sims, non-monetarists endow a passive role on money and emphasize the causal arrow running from asset prices to money demand. Sims claimed that the non-monetarist explanation effectively captures the postwar dynamic dependence between money stock and income. Sims summarized the two views as follows:

(a.1) From the non-monetarists' perspective, the expectation of future profitability causes present asset prices. An increase in asset prices increases money demand and income, whereas interest rates and asset prices are statistically associated.
(a.2) Monetarists believe that a causal relation runs from money to income, ignoring the confounding effects of asset prices.

- Monetarist explanation:

- Non-monetarist explanation:

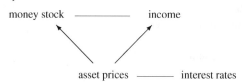

Fig. 4.10 Two Graphs of Causality

The views are graphically represented in Fig. 4.10.

The empirical analysis in Geweke (1984) addressed monthly US data from January 1966 to December 1979 on the rate of return for Treasury Bills (r), the growth rate of industrial production (y), and the growth rate of the money supply (m). Exhibiting the estimates of the frequency-wise and overall measures of conditional linear feedback and the overall conditional instantaneous feedback, Geweke concluded the following.

(b.1) Overall, the empirical results indicate that the linear feedback effect from m to y conditional on r is stronger than the one from r to y conditional on m.

(b.2) Both r and m contribute to y; however, r plays a dominant role in explaining the cyclical fluctuations in y. At frequencies corresponding to the business cycle, the conditional and unconditional linear feedbacks from r are more important than that from m. Namely, at periodicities ranging from three to five years, r accounts for more than half of the variance in y, whereas m accounts for 30%.

4.3.1.2   Predictive Content of Macroeconomic Series

According to Stock and Watson (2003), the main concepts of macroeconomics are framed using the Irving Fisher's hypotheses.

(c.1) The nominal interest rate is the real rate plus the expected inflation rate.

(c.2) A monetary contraction temporarily produces high interest rates (an inverted yield curve) and leads to an economic slowdown.

(c.3) Stock prices reflect the expected present discounted value of future earnings.

A significant number of empirical macroeconomic studies has addressed the predictive ability of the term structure and other asset-price characteristics for the future growth rate of economic activities and inflation rates. Stock and Watson (2003) and Wheelock and Wohar (2009) supplied wide-ranging reviews of the literature. Notably, much of the empirical analysis in the literature, including Hamilton and Kim (2002), is based on the ordinary regression model or the scalar ARDL model for relating the GDP growth rate and the term spread, but does not use the vector ARMA model.

Staiger, Stock and Watson (1997) investigated the predictive ability of the unemployment rate in predicting the inflation rate and concluded that it is not a useful predictor. The empirical failure of Freedman's Phillips curve and the observation of the inability of output gap for inflation prediction seem to have led Stock and Watson (2003) to focus on the relations of asset prices to output or inflation. Kim and Limpaphayom (1997) examined the effect of economic regimes on the relation between the term structure of interest rates and future economic activity in Japan. They concluded that the term structure successfully predicts future economic growth only during 1984–1991, which was the period of financial market liberalization and interest rate deregulation.

Stock and Watson (2003) posed empirical questions mainly in the framework of the single-equation ARDL model of the form

$$x(t) = \sum_{j=1}^{a} \alpha[j]x(t-j) + \sum_{k=1}^{b} \beta[k]y(t-k) + \varepsilon(t) \tag{4.11}$$

and asked the utility of the inclusion of $\sum_{k=1}^{b} \beta[k]y(t-k)$. The main conclusions of their paper are as follows:

(d.1) Some asset prices have substantial marginal predictive content for output and/or inflation at some times and in some countries. The evidence that asset prices are useful for forecasting output growth is stronger than for inflation. In the USA, there was a structural (parameter) change in the US macroeconomy around the middle of the 1980s.

(d.2) Conventional in-sample Granger causality tests only provide a poor guide for forecasting performance. Econometric methods for identifying a potentially useful predictor rely ordinarily on in-sample significance tests, such as Granger causality tests, but doing so provides little assurance that the identified predictive relation is stable.

(d.3) There is considerable instability in bivariate and trivariate predictive relations involving asset prices or other predictors. Simple methods of combining information on various predictors circumvent the worst of these instability problems.

Breitung and Candelon (2006) applied their testing method (alluded to in Remarks 4.2 and 4.3) to US quarterly GDP data, the term spread (i.e., the difference between the yields on long-term and short-term Treasury securities), and the real balance for the sample period 1959I–1998IV and investigated the predictive content of the term spread for output growth. They found a significant contribution around the periodicities of one year and the business cycle. Gronwald (2009) employed the VAR model and German macro and financial monthly data for the sample period 1963:1 through 2006:8 and, using the testing method of Breitung and Candelon, found that (1) short-run causality exists between oil prices and variables such as short-term interest rates and the German share price index; (2) there is a long-run causality between oil prices and long-term interest rates; (3) oil prices predict the consumer price index at a large number of different frequencies; and (4) there is no significant causality from oil prices to industrial production and the unemployment rate.

Assenmacher-Wesche, Gerlach, and Sekine (2008) investigated frequency-wise causal characterizations of quantity-theoretic relations using quarterly Japanese data for the CPI, M2+CD, real GDP, and call rates from 1970I to 2005IV. Using the ARDL model, they conducted statistical inferences on the causal relations among inflation, money growth, output growth, interest rates, and the output gap constructed from the data set. Additionally, they conducted inferences on partial measures between money growth and inflation. They made the following findings.

(e.1) Money growth and real output cause inflation at low frequencies, whereas the output gap causes inflation at higher frequencies. Specifically, on the low-frequency band corresponding to longer than four-year periodicity, money growth is correlated with inflation, reflecting the one-way Granger causality. In contrast, those quantity-theoretic characteristics are not observed on the high-frequency band in which the output gap is observed to one-sidedly cause inflation.

(e.2) The causality from money growth to inflation is a trait on the low-frequency band, and its stability over time cannot be confirmed empirically.

(e.3) Those quantity-theoretic traits found in Japan apply to neither the Eurozone nor the USA.

Wheelock and Wohar (2009) surveyed the literature on the usefulness of the term spread for predicting changes in economic activity, particularly for predicting changes in output growth, inflation, industrial production, consumption, and recessions. By focusing not on the causality issue itself but on predictability, they note the considerable variation in the prediction ability of the term spread across countries and over time in predicting changes in a variety of economic activities.

4.3.1.3   Discussion

Sims' "innovation accounting" is a method for identifying the intrinsic innovation component of a variable that causes another variable in view of reduced MA representation (see (3.11) for the covariance matrix-diagonalized MA representation). However, Sims used element-wise diagonalization instead of block-wise diagonalization. Remark 3.2 of Chap. 3 compares the method of this book with that of Sims.

Sharing the criticism by Granger (1997) that conventional Granger causality tests are too concerned with testing in-sample goodness of fit and not with the primary purpose of testing out-sample prediction content, Stock and Watson (2003) valued the out-sample prediction ability of a predictor rather than testing its in-sample significance. Regarding the previous point (d.1), they do not seem interested in a conditional prediction in the presence of a third series. The same suggestion as (d.3) is given in Granger (1999). Regarding (d.2), we need to note that the Granger test of significance does not question the degree to which the prediction is improved by including the sum $\sum_{k=1}^{b} \beta[k] y(t - k)$ if the null hypothesis $(\beta[1], \ldots, \beta[b]) = 0$

is rejected and whether the relation (4.11) is stable and extends to the out-sample period. The main reason for the negative assessment of Stock and Watson (2003) of the Granger causality test with respect to the prediction ability is that the in-sample significance of a test statistic of the coefficients of interest does not imply the usefulness of the corresponding candidate predictor because statistical significance does not by itself imply the out-sample stability of the relation. However, the point is not a proper weakness of the Granger non-causality test of significance. If a relation changes over time, we may expect that the in-sample observation is not extrapolated for out-sample prediction.

Finally, although Stock and Watson do not allude to confidence statements, they seem fit to express the strength of the effects. By quantifying various aspects of causality from one series to another, the approach by Geweke (1984) and this book enable the confidence set statement on interdependency measures (see Yao and Hosoya 2000).

To take into account possible structural changes in the US economy, our empirical analysis in the next section investigates how the choice of observation periods affects estimating and testing the results of the measures. Moreover, Chap. 5 provides a more explicit discussion of the structural change problem.

4.3.2 Application of the Partial Measures to US Macroeconomic Data

In this section, the proposed approach to partial measures of interdependence is applied to the empirical analysis on the relation between the term spread and the growth rate of the economy based on a set of US macroeconomic quarterly series (whose source is the Saint Louis Federal Reserve Bank database). Specifically, to evaluate the predictive ability of the term spread for the future growth rate, we focus on the interdependence between the annualized real GDP growth rate over the next quarter, which is defined by $\triangle GDP(t) \equiv 400 \times (\log GDP(t) - \log GDP(t-1))$, and the term spread ($TS$), which is defined as the difference between the government 10-year bond yield and the 3-month bond yield. The third series that we consider in the following analysis is:

a. $\triangle M2$: annualized log-difference of real $M2$;
b. $\triangle CPI$: annualized log-difference of consumer price index.

Our full sample period of US economic data is from $1959Q1$ through $2015Q2$ (denoted by Period 1+2). To investigate the change in causal structures between $\triangle GDP$ and TS, we divide the period into two subperiods of $1959Q1$ through $1984Q4$ (Period 1) and $1985Q1$ through $2015Q2$ (Period 2). To investigate the robustness of the relations, partial measures of interdependence are estimated for three periods. We base our analysis on the following three stationary vector ARMA models:

a. Model 1: Bivariate model for $\triangle GDP$, TS;

Table 4.4 Selected lag orders by AIC

Model	Period 1 (up to 1984Q4)	Period 2 (from 1985Q1)	Period 1+2 (full)
Model 1: Simple	$(1, 2)$	$(1, 1)$	$(5, 1)$
Model 2: Partial ($\triangle M2$)	$(1, 1)$	$(1, 1)$	$(2, 2)$
Model 3: Partial ($\triangle M2, \triangle CPI$)	$(2, 0)$	$(2, 0)$	$(3, 0)$

  b. Model 2: Trivariate model for $\triangle GDP$, TS, $\triangle M2$;

  c. Model 3: Four-variate model for $\triangle GDP$, TS, $\triangle CPI$, $\triangle M2$.

For all of the models and data sets, the maximum ARMA lag orders are set as $(5, 5)$ for Models 1 and 2, and $(3, 3)$ for Model 3, respectively. The selected lag orders of ARMA by AIC for the respective models and periods are listed in Table 4.4

  For brevity, we use the following notations:

- OMO($\triangle M2$) $\equiv$ Overall partial measure of one-way effect ($TS \rightarrow \triangle GDP|\triangle M2$)
- OMO2($\triangle M2$) $\equiv$ OMO($\triangle GDP \rightarrow TS|\triangle M2$)
- FMO($\triangle M2$) $\equiv$ Frequency-wise partial measure of one-way effect
$$(TS \rightarrow \triangle GDP|\triangle M2)$$
- FMO2($\triangle M2$) $\equiv$ FMO($\triangle GDP \rightarrow TS|\triangle M2$)
- OW($\triangle M2$) $\equiv$ Overall partial Wald statistic for testing ($TS \nrightarrow \triangle GDP|\triangle M2$)
- OW2($\triangle M2$) $\equiv$ OW($\triangle GDP \nrightarrow TS|\triangle M2$)
- FW($\triangle M2$) $\equiv$ Frequency-wise partial Wald statistic for testing ($TS \nrightarrow \triangle GDP|\triangle M2$)
- FW2($\triangle M2$) $\equiv$ FW($\triangle GDP \nrightarrow TS|\triangle M2$)

If the third series is evident by context, the argument is omitted such that, for instance, FW2($\triangle M2$) may be sometimes written as FW2. In the sequel, the overall and frequency-wise Wald statistics are constructed for the null hypotheses $\Gamma_{12}^{\dagger}(z) = 0$ in (3.9) and (4.4), respectively.

Remark 4.4 Before estimating the VARMA model, we conducted unit-root tests to determine the order of integration of the four series involved. The ADF, PP, and KPSS unit-root tests are used. The test results for Period 1+2 suggest that both $\triangle GDP$ and TS are stationary series, and log CPI and log $M2$ are unit-root $I(1)$ series.

(1) Simple relation between $\triangle GDP$ and TS (Model 1)

Tables 4.5, 4.7, and 4.8 list the values of the overall Wald statistics, the p-values, and the estimates for the three models, respectively. Regarding the overall one-way effects limited to the simple relation between $\triangle GDP$ and TS, Table 4.5 shows that OW for

Table 4.5 Wald statistics and overall measure estimates for Model 1

	Statistic	DF	p-value	Estimate
Simple: Period 1 (up to 1984Q4)				
OW	3.000	3	0.392	0.127
OW2	0.214	3	0.975	0.087
Simple: Period 2 (from 1985Q1)				
OW	2.278	2	0.320	0.034
OW2	1.981	2	0.371	0.118
Simple: Period 1+2 (full)				
OW	54.446	6	0.000	0.047
OW2	4.520	6	0.607	0.147

Note Estimate denotes OMO or OMO2 estimate

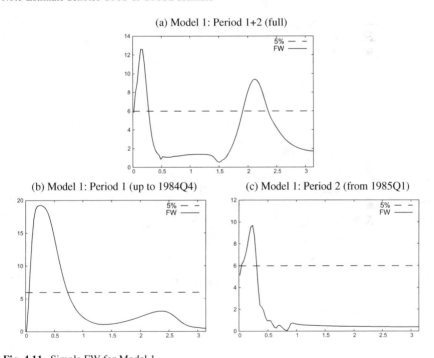

(a) Model 1: Period 1+2 (full)

(b) Model 1: Period 1 (up to 1984Q4)   (c) Model 1: Period 2 (from 1985Q1)

Fig. 4.11 Simple FW for Model 1

Period 1+2 indicates a definitive significance of *TS* simply causing $\triangle GDP$, and there is no other implication for this period. For both Period 1 and Period 2, no significant one-way effect in either direction is observed. Figure 4.11 exhibits that FW has a significant peak for each period, whereas FW2 in Fig. 4.12 is, characteristically, not significant on the entire frequency domain for any of the three periods—consistent with the results of the OW test. Although the OW test results for Periods 1 and 2 do not reject the null hypothesis, there exists a significant frequency band in which

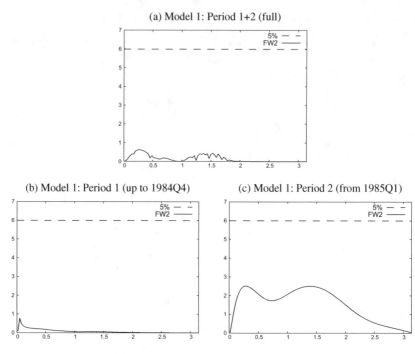

Fig. 4.12 Simple FW2 for Model 1

Table 4.6 Correspondence between frequency and periodicity

Frequency	0	$\pi/8$	$\pi/4$	$3\pi/8$	$\pi/2$	$5\pi/8$	$7\pi/8$	π
Periodicity (quarters)	∞	16	8	16/3	4	16/5	16/7	2

TS simply causes $\triangle GDP$ for any period. Specifically, for Periods 1 and 2, FW has a notable mass around $\lambda = 0.08\pi$ and $\lambda = 0.07\pi$, respectively, whereas FW has two significant peaks for Period 1+2; one is near $\lambda = 0.05\pi$ and the other is a neighborhood of $\lambda = 0.63\pi$. Because we use quarterly data, they imply that there exist a six-year and one-quarter periodicity for Period 1 and an approximate seven-year periodic cycle for Period 2. Moreover, for Period 1+2, one is a 10-year periodicity and the other is a three-quarter periodicity. The relationship between the frequency and the periodicity is displayed in Table 4.6.

Once significant bands are identified, the next step is to evaluate the strength of the frequency-wise one-way effect, which is given in Fig. 4.13. For Periods 1 and 2, Figs. 4.13b and c show that the partial measures of interdependence appear very similar to each other. Although FMO2 diverges at the origin $\lambda = 0$, it is very small on the entire domain. Regarding Period 1+2, Fig. 4.13a indicates that FMO and FMO2 have two peaks, respectively; however, FMO2 is not significant on the entire frequency domain as observed in Fig. 4.12a. Regarding the simple measure of the

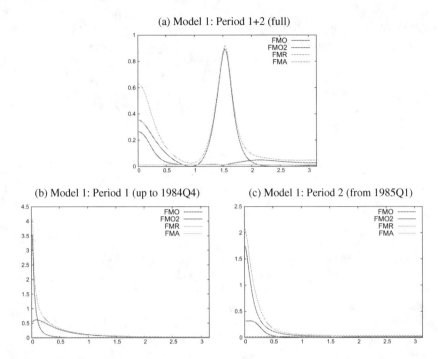

Fig. 4.13 Simple measures for Model 1

one-way effect from TS to $\triangle GDP$, Figs. 4.13a and 4.11a show that two peaks in FMO are statistically significant at the 5% level, and the one-way effect from TS to $\triangle GDP$ on the low-frequency domain is stronger than the one on the high-frequency domain. The maximum effects are 0.265 and 0.048, respectively. We find that the one-way effect from TS to $\triangle GDP$ in six-year to 50-year periodicities is approximately six times stronger than that of an approximate three-quarter periodicity.

A comparison of Figs. 4.13b with c indicate the robustness of the foregoing observations. Even if the whole sample period is divided into two subperiods, we still find one peak of FMO on the low-frequency domain for Periods 1 and 2, whereas the peak in a short periodicity vanishes, which is not significant, as in Figs. 4.11b and c. For both periods, the peak is included on the significant frequency band. The maximum sizes of the one-way effects from TS to $\triangle GDP$ for Periods 1 and 2 are 0.620 and 0.318, respectively. After the middle of the 1980s, the relative impact of the simple one-way effect from TS to $\triangle GDP$ becomes approximately one-half.

(2) Partial relation between $\triangle GDP$ and TS in the presence of the third series $\triangle M2$ (Model 2)

The overall Wald statistic values in Table 4.7 indicate that there is no significant one-way effect observable in either direction, except for OW for Period 1 for which significant causality is observed, with the one-way effect estimate of 0.275. In con-

Table 4.7 Wald statistics and overall measure estimates for Model 2

	Statistic	DF	p-value	Estimate
Partial: Period 1 (up to 1984Q4)				
OW($\triangle M2$)	8.231	3	0.041	0.275
OW2($\triangle M2$)	0.193	3	0.979	0.012
Partial: Period 2 (from 1985Q1)				
OW($\triangle M2$)	1.920	3	0.589	0.030
OW2($\triangle M2$)	2.230	3	0.526	0.132
Partial: Period 1+2 (full)				
OW($\triangle M2$)	7.278	6	0.296	0.058
OW2($\triangle M2$)	3.474	6	0.747	0.081

Note Estimate denotes OMO or OMO2 estimate

trast, in terms of the frequency-wise non-causality test for Period 1, Figs. 4.14b and 4.15b show that FW has a significant spire-like peak near the origin and two significant bands, such as $[0.01\pi, 0.05\pi)$ and $(0.05\pi, 0.46\pi]$, whereas FW2 stays very low and non-significant on the entire frequency domain.

For Period 2, Figs. 4.14c and 4.15c indicate that FW has a significant frequency band below $\lambda = 0.14\pi$, but FW2 has no significant frequency band as in Period 1. However, for Period 2, FMO2 is ascendant on the entire domain, as shown in Fig. 4.16c.

For Period 1+2, the FW statistic has two peaks around $\lambda = 0.06\pi$ and 0.3π, respectively. However, both have very narrow significant bands, such as $[0.06\pi, 0.08\pi]$ and $[0.28\pi, 0.32\pi]$ reaching the 5% significance level. FW2 has a sharp and significant peak at $\lambda = 0.04\pi$, which is a feature quite different from the one in the case of simple FW2. Unlike Model 1, Fig. 4.16a shows that for Period 1+2, FMO and FMO2 have a single peak at 0.06π and around the origin, and FMO2 dominates FMO for $\lambda < 0.06\pi$, whereas FMO dominates FMO2 otherwise. Figure 4.16b shows that FMO dominates FMO2, and FMO2 maintains a very low level on the entire frequency domain for Period 1, whereas for Period 2 FMO2 maintains a higher level than FMO on the entire frequency domain.

Taking into account the significant frequency bands, we observe that for Period 1+2, the one-way effect from TS to $\triangle GDP$ is 0.400 at the peak level of approximately an eight-year and one-quarter periodicity, and the opposite direction of the one-way effect is 0.851 at the peak level, which is two times higher than the one-way effect from TS to $\triangle GDP$. However, the periodicity is longer. The significant frequency bands for FMO and FMO2 on the low frequency are from a six-and-half-year to a nine-year periodicity for FMO and the periodicity exceeding approximately a four-and-half-year for FMO2, respectively. Comparing Periods 1 and 2, we find that the one-way effect from TS to $\triangle GDP$ is relatively weakened after the middle of the 1980s, whereas FW2 is not significant on the entire frequency domain for both of the periods.

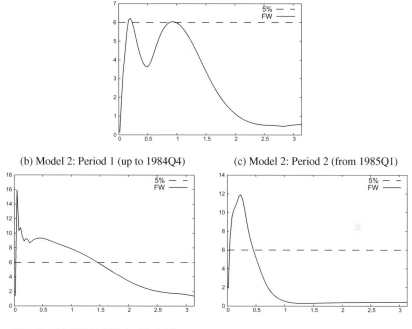

Fig. 4.14 Partial FW ($\triangle M2$) for Model 2

Table 4.8 Wald statistics and overall measure estimates for Model 3

	Statistic	DF	p-value	Estimate
Partial: Period 1 (up to 1984Q4)				
OW($\triangle M2$, $\triangle CPI$)	19.315	6	0.004	0.257
OW2($\triangle M2$, $\triangle CPI$)	3.421	6	0.754	0.019
Partial: Period 2 (from 1985Q1)				
OW($\triangle M2$, $\triangle CPI$)	0.963	6	0.987	0.006
OW2($\triangle M2$, $\triangle CPI$)	11.343	6	0.078	0.106
Partial: Period 1+2 (full)				
OW($\triangle M2$, $\triangle CPI$)	14.762	9	0.098	0.057
OW2($\triangle M2$, $\triangle CPI$)	14.280	9	0.113	0.055

Note Estimate denotes OMO or OMO2 estimate

(3) Partial relation between $\triangle GDP$ and TS in the presence of $\triangle M2$ and $\triangle CPI$ (Model 3)

Table 4.8 shows that both OW and OW2 are weakly significant for Period 1+2, and estimates of OMO and OMO2 are given by 0.057 and 0.055, respectively. This might indicate that FMO and FMO2 are very similar to each other. However, characteristically, there is a marked contrast between the two periods; for Period 1, only OW

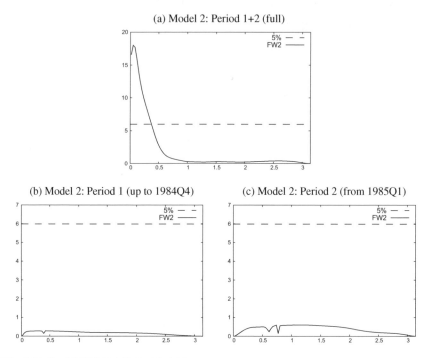

Fig. 4.15 Partial FW2($\triangle M2$) for Model 2

is significant, whereas for Period 2, only OW2 is significant. Namely, the direction of causality changed drastically. Estimates of OMO for Period 1 and OMO2 for Period 2 are 0.257 and 0.106, respectively. By eliminating the third series effects of $\triangle CPI$ and $\triangle M2$, the result seems in accordance with the assertion of the literature on the decline of the prediction ability of TS after the middle of the 1980s (see Sect. 4.3.1.2). In addition, these overall results are consistent with the frequency-wise test results. From Figs. 4.17b and 4.18b, we see that for Period 1 there is a sizeable band of significance on the low-frequency domain $(0.04\pi, 0.51\pi)$, whereas FW2 is not significant on the entire domain. For Period 2, Figs. 4.17c and 4.18c show that FW is not significant on the entire domain, whereas FW2 is significant on the band $(0.02\pi, 0.48\pi)$. For Period 1+2, in Figs. 4.17a and 4.18a, both statistics of FW and FW2 have sizeable bands, such as $(0.09\pi, 0.35\pi)$ and $(0.01\pi, 0.51\pi)$, respectively, where the statistics are significant.

For all three periods, estimates of the partial one-way effect measures in Fig. 4.19 are very similar to the ones in Fig. 4.16. The exceptions are, in Fig. 4.19a, FMO dominating FMO2 for $\lambda < 0.11\pi$ and FMO2 dominating FMO otherwise, whereas Fig. 4.16a shows that FMO2 dominates FMO if $\lambda < 0.06\pi$, but that ascendancy is reversed otherwise. We divide the sample period into two subperiods and find on the whole that the one-way effect from TS to $\triangle GDP$ is significant and dominant for Period 1, whereas the opposite direction of causality is observed significantly

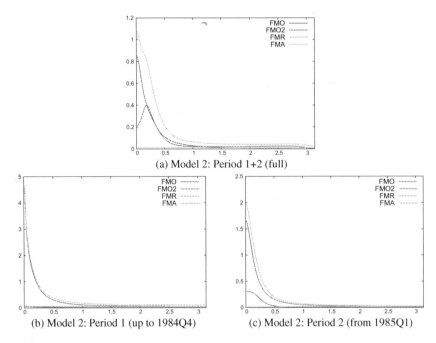

(a) Model 2: Period 1+2 (full)

(b) Model 2: Period 1 (up to 1984Q4)   (c) Model 2: Period 2 (from 1985Q1)

Fig. 4.16 Partial measures($\triangle M2$) for Model 2

for Period 2. By adding $\triangle CPI$ as one of the third series, the strengths of the partial measures are somewhat weakened; however, the contrast in the related Wald statistics before and after the middle of the 1980s becomes clearer. The pictures given by the simple measures of the one-way effect in Fig. 4.13 fail to exhibit the alternation of dominance.

The proposed analysis of causality consists of the following steps:

(1) To evaluate the overall Wald statistics and the strength of the overall measures when the statistics are significant. Even if the null hypothesis of non-causality for an overall measure is accepted, frequency-wise causality is possible.
(2) To evaluate the frequency-wise Wald statistics and the strengths of the frequency-wise measures.
(3) For examining the robustness of the significance, strength, and direction of the measures, to consider different sets of the third series and divide the original sample period into subperiods.
(4) To apply structural change tests for measures as expounded in Chap. 5.

Our approach to a frequency-domain causal analysis helps the elucidation of a detailed causal structure among multivariate time-series data. The proposed approach has the merit of applicability to general time-series models as long as they have numerically factorizable spectral density. Moreover, the inference is not limited to testing Granger non-causality, but, as shown in the preceding arguments, it is applica-

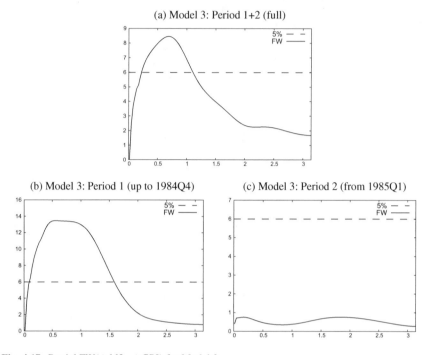

Fig. 4.17 Partial FW($\triangle M2$, $\triangle CPI$) for Model 3

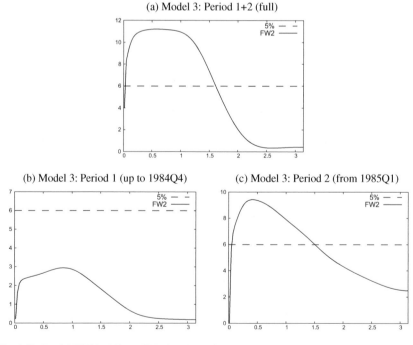

Fig. 4.18 Partial FW2($\triangle M2$, $\triangle CPI$) for Model 3

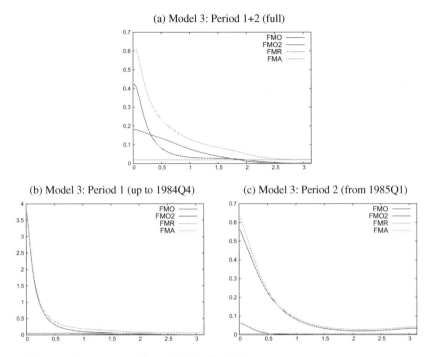

Fig. 4.19   Partial measures($\triangle M2$, $\triangle CPI$) for Model 3

ble to a broader range of the quantitative analysis of interdependence. For further
development, possible topics include:

- Estimation method for better finite sample performance with conformity of large-sample plausibility;
- Extension of the empirical analysis to long-range-dependent or non-stationary-cointegrated models; the latter extension was partially given by Yao and Hosoya (2000);
- A rewrite of the entire approach in terms of nonparametric time-series statistics; and
- Development of statistical causal analysis conformable with out-sample predictions.

References

Assenmacher-Wesche, K., Gerlach, S., & Sekine, T. (2008). Monetary factors and inflation in Japan. *Journal of the Japanese and International Economies*, *22*, 343–363.

Breitung, J., & Candelon, B. (2006). Testing for short- and long-run causality: A frequency-domain approach. *Journal of Econometrics*, *132*, 363–378.

Dufour, J.M., & Pelletier, D. (2011). Practical methods for modelling weak VARMA prcoesses: Identification, estimation and specialization with macroeconomic application, *working paper*.

Durbin, J. (1960). The fitting of time-series models. *International Statistical Review*, *33*, 233–244.

Geweke, J. (1984). Measures of conditional linear dependence and feedback between time series. *Journal of the American Statistical Association*, *79*, 907–915.

Granger, C. W. J. (1997). The ET interview: Professor Clive Granger. *Econometric Theory*, *13*, 253–303.

Granger, C. W. J. (1999). *Empirical Modeling in Economics: Specification and Evaluation*. Cambridge: Cambridge University Press.

Gronwald, M. (2009). Reconsidering the macroeconomics of the oil price in Germany: Testing for causality in the frequency domain. *Empirical Economics*, *36*, 441–453.

Hamilton, J. D., & Kim, D. H. (2002). A reexamination of the predictability of economic activity using the yield spread. *Journal of Money, Credit and Banking*, *34*, 340–360.

Hannan, E. J., & Rissanen, J. (1982). Recursive estimation of mixed autoregressive-moving average order. *Biometrika*, *69*, 81–94.

Hannan, E. J., & Kavalieris, L. (1984a). A method for autoregressive-moving average estimation. *Biometrika*, *71*, 273–280.

Hannan, E. J., & Kavalieris, L. (1984b). Multivarate linear time series models. *Advances of Applied Probability*, *16*, 492–561.

Hosoya, Y. (1997). A limit theory for long-range dependence and statistical inference on related models. *The Annals of Statistics*, *25*, 105–137.

Hosoya, T., & Takimoto, T. (2010). A numerical method for factorizing the rational spectral density matrix. *Journal of Time Series Analysis*, *31*, 229–240.

Johansen, S. (1995). *Likelihood-based Inference in Cointegrated Autoregressive Models*. Oxford: Oxford University Press.

Judd, K. L. (1999). *Numerical Methods in Economics*. Cambridge: The MIT Press.

Kim, K. A., & Limpaphayom, P. (1997). The effect of economic regimes on the relation between term structures and real activity in Japan. *Journal of Economic and Business*, *49*, 379–392.

Sims, C. A. (1980). Comparison of interwar and postwar business cycles: Monetarism reconsidered. *The American Economic Review*, *70*, 250–257.

Staiger, D., Stock, J. H., & Watson, M. W. (1997). The NAIRU, unemployment and monetary policy. *Journal of Economic Perspective*, *11*, 33–49.

Stock, J. H., & Watson, M. W. (2003). Forecasting output and inflation: The role of asset prices, *Journal of Economic Literature*, *XLI*, 788–829.

Takimoto, T., & Hosoya, Y. (2004). A three-step procedure for estimating and testing cointegrated ARMAX models. *The Japanese Economic Review*, *55*, 418–450.

Takimoto, T., & Hosoya, Y. (2006). Inference on the cointegration rank and a procedure for VARMA root-modification. *Journal of Japan Statistical Society*, *36*, 149–171.

Wheelock, D. C., & Wohar, M. E. (2009). Can the term spread predict output growth and recession? *Federal Reserve Bank of St. Louis Review*, September/October, *Part 1*, 419–440.

Whittle, P. (1952). Some results in time series analysis. *Skandinavisk Aktuarietidskrift*, *1–2*, 48–60.

Whittle, P. (1953). The analysis of multiple stationary time series. *Journal of the Royal Statistical Society B*, *15*, 125–139.

Yao, F., & Hosoya, Y. (2000). Inference on one-way effect and evidence in Japanese macroeconomic data. *Journal of Econometrics*, *98*, 225–255.

Yap, S. F., & Reinsel, G. C. (1995). Estimation and testing for unit roots in a partially nonstationary vector autoregressive moving average model. *Journal of the American Statistical Association*, *90*, 253–267.

Chapter 5
Inference on Changes in Interdependence Measures

Abstract The causal relationship between the time series can be characterized with the moments of distributions for the series and the parameters of models such as the vector ARMA model from previous chapters. Thus, the changes in the moments of the time series and the model parameters suggest the possibility of a change in causal relationships as we expected. However, the changes in the moments and the model parameters do not tell us much about the magnitude of the change in causal relationships. In this chapter, we provide a measure of the change in causal relationships between a time series and the test statistic to determine whether such a change is associated with a structural change and is statistically significant. The properties of the measure and the test statistic are examined through a Monte Carlo simulation, and empirical examples are provided.

Keywords Change in measure · High-frequency data · Subsampling · Variance estimation

5.1 Change in Measures

An economic time series often has a structural change ascribed to a financial crisis and a policy change. The time series for the periods before and after a structural break may be characterized by models with different parameters. To detect a structural change and to identify a break date, several methods proposed in previous studies, e.g., Andrews (1993) and Hansen (1996), are applicable.

Although the presence of a structural change is detected by changes in model parameters, these changes do not reveal much about how a causal relationship changes. In other words, it is not obvious how causality changes even when large changes in parameters are confirmed. The following example illustrates the case that there is no change in partial causality directed from $\{y(t)\}$ to $\{x(t)\}$ despite the model parameters being changed.

© The Author(s) 2017
Y. Hosoya et al., *Characterizing Interdependencies of Multiple Time Series*,
JSS Research Series in Statistics, DOI 10.1007/978-981-10-6436-4_5

Example 5.1 VAR(2) process with structural break

$$\begin{cases} x(t) = 0.2y(t-1) + 0.5z(t-1) - (0.25 + 0.05 \cdot 1_{t>T_1})y(t-2) + \varepsilon_1(t), \\ y(t) = \varepsilon_2(t), \\ z(t) = (0.5 + 0.1 \cdot 1_{t>T_1})y(t-1) + \varepsilon_3(t) \end{cases}$$

where $\{\varepsilon_1(t)\}$, $\{\varepsilon_2(t)\}$, and $\{\varepsilon_3(t)\}$ are mutually independent white noise processes. The absolute value of the coefficient of $y(t-2)$ in the first equation for $x(t)$ becomes large after the break date T_1. Seemingly, the causality directed from $\{y(t)\}$ to $\{x(t)\}$ is strengthened. However, the partial causality from $\{y(t)\}$ to $\{x(t)\}$ is unchanged because $z_{0,0,-1}(t) = \varepsilon_3(t)$, $u(t) = \varepsilon_1(t) + 0.2\varepsilon_2(t-1)$, and $v(t) = \varepsilon_2(t)$ for all sample periods. For details on the variables $z_{0,0,-1}(t)$, $u(t)$, and $v(t)$, see Sect. 3.2 and the examples therein.

Because the overall measure is represented by the integral of the frequency-wise measure, as shown in (2.17), it is possible that there is no change in the overall measures but there is some change in the frequency-wise measures. Although the example subsequently provided is artificial, it illustrates the described case.

Example 5.2 VMA(2) process with structural break

$$\begin{cases} x(t) = \varepsilon_1(t) + (-0.25 + 0.5 \cdot 1_{t>T_1})\varepsilon_1(t-1) + 0.1\varepsilon_2(t-2), \\ y(t) = \varepsilon_2(t) \end{cases}$$

where $\{\varepsilon_1(t)\}$ and $\{\varepsilon_2(t)\}$ are mutually independent white noise. Denote the causality measures at frequency λ for the periods before and after a break as $M_b(\lambda)$ and $M_a(\lambda)$, respectively. Then, $M_b(\lambda) \neq M_a(\lambda)$ for $\lambda \in (0, \pi)$ except $\lambda = \pi/2$ in contrast to the overall measure, which does not change across a break because $M_b(\pi/2 - \lambda) = M_a(\pi/2 + \lambda)$ holds for $\lambda \in (-\pi/2, \pi/2)$.

These examples arouse the necessity to evaluate a change in the frequency-wise measure itself. In the following, we introduce a statistical inference for a change in a causality relationship for the stationary vector ARMA model. Because many ingredients have been provided in previous chapters, we mainly focus on the essentials to help understand how to measure a change in a causal relationship.

5.1.1 Change in Measures for Stationary Vector ARMA Model

We consider the case with one possible structural break point in the entire sample period and divide the entire sample into two subsamples before and after the break point. Let T and T_1 be an entire sample size and a sample size of the period before a break point. The sample size of the period after a break is given as $T_2 = T - T_1$.

Assume that the break point is a known constant fraction of T, i.e., $T_1 = [cT]$, where $[\cdot]$ stands for a Gauss symbol. In fact, T_1 is the end point of the subsample before a break and the break point is not T_1 but is $T_1 + 1$. Because the difference in the definition of a break point is ignorable asymptotically, we use T_1 as a break point in what follows.

Consider the following vector process $\{w(t)\}$ that is generated by a stationary multivariate ARMA process with a structural break

$$A_k(L)w(t) = B_k(L)\varepsilon(t), \quad w(t) = \begin{bmatrix} x(t) \\ y(t) \\ z(t) \end{bmatrix}, \quad \varepsilon(t) = \begin{bmatrix} \varepsilon_1(t) \\ \varepsilon_2(t) \\ \varepsilon_3(t) \end{bmatrix}, \quad t \in \mathbb{Z} \quad (5.1)$$

where $x(t)$, $y(t)$, and $z(t)$ are p_1, p_2, and p_3 dimensional random vectors and the innovation $\{\varepsilon(t)\}$ is an *i.i.d.* white noise process with zero mean and covariance matrix $\Sigma_k^\dagger$. The subscript k takes the value of 1 if $t \leq T_1$ and 2 otherwise. $A_k(L) = \sum_{j=0}^{a} A_k[j]L^j$ and $B_k(L) = \sum_{j=0}^{b} B_k[j]L^j$ are ath and bth order polynomials of the lag operator L with $A_k[0] = B_k[0] = I_p$ and $p = p_1 + p_2 + p_3$. We employ the same assumptions for $A_k(z)$, $B_k(z)$, and the spectral density for the process $\{w(t)\}$, as in Chap. 3. For notational simplicity, we express the orders of the lag polynomials $A_k(L)$ and $B_k(L)$ as a and b, although these may depend on k.

Let θ_k be a $n_{\theta,k}$-vector parameter and assume that the model parameter of (5.1) is a function of θ_k, as we have seen in Chap. 4. The estimator of the partial interdependence measure between $\{x(t)\}$ and $\{y(t)\}$ in the presence of $\{z(t)\}$ introduced in the previous chapters inherits the statistical properties from the parameter estimator $\hat{\theta}_k$. The estimation method introduced in Sect. 4.1.1 can be applied to each model for the periods before and after a structural break. The asymptotic properties of the estimators are also provided in Sect. 4.1.3. See Hosoya (1997) for details.

Let $PM_{y \to x:z}(\theta_k, \lambda)$ be a partial frequency-wise measure of the one-way effect from $\{y(t)\}$ to $\{x(t)\}$ in the presence of $\{z(t)\}$ at frequency λ, and let $PM_{y \to x:z}(\hat{\theta}_k, \lambda)$ be its estimator.

Under the assumption of $PM_{y \to x:z}(\theta_k, \lambda) > 0$ in a neighborhood of the true θ_k, the asymptotic distribution of the partial measure estimator $PM_{y \to x:z}(\hat{\theta}_k, \lambda)$ is shown to be

$$\sqrt{T_k}(PM_{y \to x:z}(\hat{\theta}_k, \lambda) - PM_{y \to x:z}(\theta_k, \lambda)) \xrightarrow{d} N(0, H(\theta_k, \lambda)) \quad (5.2)$$

based on the asymptotic normality of the parameter estimator $\hat{\theta}_k$, that is $\sqrt{T_k}(\hat{\theta}_k - \theta_k) \xrightarrow{d} N(0, \Psi(\theta_k))$. The asymptotic variance $H(\theta_k, \lambda)$ of the partial measure estimator appearing in (5.2) is given as

$$H(\theta_k, \lambda) = D_{\theta_k} PM_{y \to x:z}(\theta_k, \lambda) \Psi(\theta_k) D_{\theta_k} PM_{y \to x:z}(\theta_k, \lambda)^*$$

where $D_{\theta_k} PM_{y \to x:z}(\theta_k, \lambda)$ is the Jacobian vector of $PM_{y \to x:z}(\theta_k, \lambda)$ evaluated at θ_k. The estimate of the asymptotic variance $H(\theta_k, \lambda)$ is available using the procedures described in Sect. 4.1.3.

The change in frequency-wise partial measures between the periods before and after a break at frequency λ can be defined as

$$CPM_{y \to x:z}(\theta_1, \theta_2, \lambda) = PM_{y \to x:z}(\theta_1, \lambda) - PM_{y \to x:z}(\theta_2, \lambda).$$

The asymptotic independence of parameter estimators $\hat{\theta}_1$ and $\hat{\theta}_2$ leads to

$$\sqrt{T}(CPM_{y \to x:z}(\hat{\theta}_1, \hat{\theta}_2, \lambda) - CPM_{y \to x:z}(\theta_1, \theta_2, \lambda)) \xrightarrow{d} N(0, V(\theta_1, \theta_2, \lambda))$$

under the assumption of $PM_{y \to x:z}(\theta_k, \lambda) > 0$ in the neighborhood of the true θ_k.

The asymptotic variance $V(\theta_1, \theta_2, \lambda)$ is represented as the linear combination of $H(\theta_1, \lambda)$ and $H(\theta_2, \lambda)$, that is

$$V(\theta_1, \theta_2, \lambda) = c^{-1} H(\theta_1, \lambda) + (1 - c)^{-1} H(\theta_2, \lambda).$$

The confidence interval of the change in the measure $CPM_{y \to x:z}(\theta_1, \theta_2, \lambda)$ with confidence coefficient $1 - \alpha$ is given as

$$\left(CPM_{y \to x:z}(\hat{\theta}_1, \hat{\theta}_2, \lambda) - z_{\alpha/2} \sqrt{\hat{h}_1 + \hat{h}_2}, \ CPM_{y \to x:z}(\hat{\theta}_1, \hat{\theta}_2, \lambda) + z_{\alpha/2} \sqrt{\hat{h}_1 + \hat{h}_2} \right)$$

where $\hat{h}_k = T_k^{-1} H(\hat{\theta}_k, \lambda)$ for $k = 1, 2$ and z_α is the upper $(1 - \alpha)$-percent point for the standard normal distribution.

To test the null hypothesis of no change in measures at frequency λ, H_0 : $CPM_{y \to x:z}(\theta_1, \theta_2, \lambda) = 0$, and the alternative H_1 : $CPM_{y \to x:z}(\theta_1, \theta_2, \lambda) \neq 0$, the test statistic can be defined as

$$CPM_{y \to x:z}(\hat{\theta}_1, \hat{\theta}_2, \lambda) / \sqrt{\hat{h}_1 + \hat{h}_2}. \tag{5.3}$$

The test statistic (5.3) is asymptotically distributed as the standard normal under H_0 and diverges to infinity under H_1 as the sample size T goes to infinity.

The change in measure of the simple measure $M_{y \to x}(\theta_k, \lambda)$, which does not take into account the third series $z(t)$ for the periods before and after the break, can be also defined as

$$CM_{y \to x}(\theta_1, \theta_2, \lambda) = M_{y \to x}(\theta_1, \lambda) - M_{y \to x}(\theta_2, \lambda).$$

The asymptotic distributions of its estimator and the related test statistic can be derived similarly to those for the partial measure change previously provided.

5.1.2 Inference for Noncausal Relationship

We have considered a change in a causal relationship only when the causal measures take positive values for both before and after a structural break. Although a change in a causal measure from zero to positive or vice versa is identified by testing for a noncausal relationship for each period, the inference for a noncausal relationship requires some additional considerations because the asymptotic normality for the measure estimator requires a positive causal measure assumption. In what follows, we provide a brief explanation of the role of the assumption and the appropriate way to address the case in which the assumption does not hold.

Denote the n_θ-vector parameter and the measure of interdependency at frequency λ as θ and $M(\theta, \lambda)$. The estimators of θ and $M(\theta, \lambda)$ are represented as $\hat{\theta}$ and $M(\hat{\theta}, \lambda)$, respectively. The first-order approximation of the estimator $M(\hat{\theta}, \lambda)$ around the true parameter θ is given as

$$\sqrt{T}\left(M(\hat{\theta}, \lambda) - M(\theta, \lambda)\right) = M^{(1)}(\theta, \lambda)\sqrt{T}(\hat{\theta} - \theta) + o_p(1)$$

where $M^{(1)}(\theta, \lambda)$ is the Jacobian vector of $M(\theta, \lambda)$ with respect to θ. The asymptotic normality of the estimator $M(\hat{\theta}, \lambda)$ comes from that of the parameter estimator $\hat{\theta}$ with the assumption of $M^{(1)}(\theta, \lambda) \neq 0$. It is noted that the measure $M(\theta, \lambda) = 0$ implies that $M(\theta, \lambda)$ takes its minimum value. Thus, a noncausal relationship, i.e., $M(\theta, \lambda) = 0$, implies $M^{(1)}(\theta, \lambda) = 0$. The first-order approximation does not provide any helpful information on the asymptotic behavior of $M(\hat{\theta}, \lambda)$ when the true measure $M(\theta, \lambda)$ equals zero.

To obtain a nondegenerated distribution of $M(\hat{\theta}, \lambda)$ for the case of $M(\theta, \lambda) = 0$, we consider the second-order approximation as follows

$$TM(\hat{\theta}, \lambda) = \frac{1}{2}\sqrt{T}(\hat{\theta} - \theta)'M^{(2)}(\theta, \lambda)\sqrt{T}(\hat{\theta} - \theta) + o_p(1)$$

$$= \frac{1}{2}\sum_{i=1}^{n_\theta}\gamma_i(\theta, \lambda)Z_i^2 + o_p(1)$$

where $M^{(2)}(\theta, \lambda)$ is the Hessian matrix of $M(\theta, \lambda)$ with respect to θ, Z_i are mutually independent standard normal random variables, $\gamma_1(\theta, \lambda) \geq \cdots \geq \gamma_{n_\theta}(\theta, \lambda)$ are the eigenvalues of matrix $\Psi(\theta)^{1/2}M^{(2)}(\theta, \lambda)\Psi(\theta)^{1/2}$, and $\Psi(\theta)$ is the asymptotic covariance matrix of $\hat{\theta}$. The asymptotic distribution of $M(\hat{\theta}, \lambda)$ is approximated by the linear combination of independent chi-square random variables with one degree of freedom, as shown in the second-order approximation.

Testing for a noncausal relationship, that is, the null of $M(\theta, \lambda) = 0$ using the approximated distribution of $M(\hat{\theta}, \lambda)$ previously described, requires numerical computations and simulations. Implementing the testing procedure is cumbersome; however, conducting the test does not take much time. In contrast, the simulation method, such as a bootstrap, is applicable to obtain the distribution of the test statistic, which is easy to implement but takes time to conduct. In most parts of this book, we use the latter method. For the other testing procedure for $M(\theta, \lambda) = 0$, Breitung and Candelon (2006) proposed a related test statistic for a simple VAR model assumption.

5.2 Tests Based on Subsampling Method

In this section, testing for a change in a causal relationship for periods before after a break point based on the subsampling method is considered. As previously described, statistical inferences require an asymptotic variance estimate of the interdependence measure estimator. The asymptotic variance estimate can be obtained using the delta method because the interdependence measure estimator is a function of the estimators of unknown parameters. The delta method requires the Jacobian of the interdependence measure, which is accompanied by numerical differentiation with respect to the parameters. Instead of the delta method, we introduce the alternative approach to provide the asymptotic variance estimate of the interdependence measure estimator. This approach does not require numerical differentiation.

5.2.1 Test for a Change in Measures Using High-Frequency Data

For an empirical analysis using a macroeconomic time series, quarterly and monthly data are typically used. In most cases, the sample size is large enough to conduct a proper statistical inference, but not large when we divide the entire sample into several subsamples. In contrast, a large sample size data set— what we call high-frequency data and big data—has become available and enables us to ensure the asymptotic inference. For example, the intra-daily data observed in a financial market are typical high-frequency data.

Suppose we have two subsample periods divided before and after a structural break, and the subsample period k consists of n_k days. The subscript $k = 1$ and 2 stands for the subsample periods before and after a structural break, respectively. Further, we assume that a sample size of intra-daily data for the ith day in subsample period k is $T_k^{(i)}$ and $T_k^{(i)} = T_k$ for simplicity.

Let θ_k and $M(\theta_k, \lambda)$ be the model parameter and the interdependence measure at frequency λ for subsample period k, respectively. The model parameter estimator and the interdependence measure estimator at frequency λ using intra-daily data for the ith day in subsample period k are denoted as $\hat{\theta}_k^{(i)}$ and $M(\hat{\theta}_k^{(i)}, \lambda)$.

The change in the interdependence measure, which is defined as $CM(\lambda) = M(\theta_1, \lambda) - M(\theta_2, \lambda)$, can be estimated using the difference in the sample averages of $M(\hat{\theta}_k^{(i)}, \lambda)$, $k = 1, 2$, that is

$$\widehat{CM}(\lambda) = \bar{M}_1(\lambda) - \bar{M}_2(\lambda) \tag{5.4}$$

where

$$\bar{M}_k(\lambda) = \frac{1}{n_k} \sum_{i=1}^{n_k} M(\hat{\theta}_k^{(i)}, \lambda). \tag{5.5}$$

The asymptotic properties for (5.4) rely on those for $M(\hat{\theta}_k^{(i)}, \lambda)$ through $\hat{\theta}_k^{(i)}$. Because properly normalized $\sqrt{T_k}(M(\hat{\theta}_k^{(i)}, \lambda) - M(\theta_k, \lambda))$ is asymptotically distributed as normal with zero mean and some positive variance, we have

$$\sqrt{n_k T_k}(\bar{M}_k(\lambda) - M(\theta_k, \lambda)) \overset{d}{\to} N(0, H_k(\lambda))$$

for the sample average (5.5) and

$$\frac{\widehat{CM}(\lambda) - CM(\lambda)}{\sqrt{\frac{H_1(\lambda)}{n_1 T_1} + \frac{H_2(\lambda)}{n_2 T_2}}} \overset{a}{\sim} N(0, 1)$$

for the change in the interdependence measure estimator. The asymptotic variance $H_k(\lambda)$ is estimated using the sample variance of $\sqrt{T_k} M(\hat{\theta}_k^{(i)}, \lambda)$ as follows

$$\hat{H}_k(\lambda) = \frac{1}{n_k} \sum_{i=1}^{n_k} T_k(M(\hat{\theta}_k^{(i)}, \lambda) - \bar{M}_k(\lambda))^2. \tag{5.6}$$

To test the null hypothesis that there is no change in measures at frequency λ, the test statistic of the null and alternative hypotheses $H_0 : CM(\lambda) = 0$ and $H_1 : CM(\lambda) \neq 0$ is given as

$$\frac{\widehat{CM}(\lambda)}{\sqrt{\frac{\hat{H}_1(\lambda)}{n_1 T_1} + \frac{\hat{H}_2(\lambda)}{n_2 T_2}}}. \tag{5.7}$$

It is easy to see that the test statistic (5.7) is asymptotically distributed as a standard normal under H_0.

In what follows, we provide a generalization of the approach to obtain the asymptotic variance estimate for the interdependence measure previously described.

5.2.2   Variance Estimation via Subsampling

Let $\{w(t)\}$ follow a stationary multivariate ARMA process, as in (4.1). Consider the case in which an entire sample $\{w(1), \ldots, w(T)\}$ is divided into nonoverlapping subsamples, $\{w(1), \ldots, w(T_b)\}$, $\{w(T_b + 1), \ldots, w(2T_b)\}$, ..., $\{w(T - T_b + 1), \ldots, w(T)\}$, where T and T_b are the size of the entire sample and the block size of the subsample under the assumption of $T = nT_b$ without loss of generality.

Denote the estimators of parameter θ on the basis of the entire sample and the ith subsample as $\hat{\theta}_T$ and $\hat{\theta}_{T_b,i}$. It is not difficult to see that $\sqrt{T}(\hat{\theta}_T - \theta)$ and $\sqrt{T_b}(\hat{\theta}_{T_b,i} - \theta)$ have the same limiting distribution $N(0, \Psi(\theta))$ under standard regularity conditions with additional assumptions $T_b/T \rightarrow 0$ and $T_b \rightarrow \infty$ as $T \rightarrow \infty$.

Suppose that $g(\theta, \lambda)$ is a differentiable positive scalar-valued function of θ and λ. Then, $\sqrt{T}(g(\hat{\theta}_T, \lambda) - g(\theta, \lambda))$ and $\sqrt{T_b}(g(\hat{\theta}_{T_b,i}, \lambda) - g(\theta, \lambda))$ are asymptotically distributed as normal with zero mean and variance $V(\theta, \lambda)$. Although the estimator of the asymptotic variance $V(\theta, \lambda)$ can be obtained using the delta method, we provide an alternative estimator using the sample variance of $\sqrt{T_b}\, g(\hat{\theta}_{T_b,i}, \lambda)$ as

$$\hat{V}(\lambda) = \frac{1}{n} \sum_{i=1}^{n} T_b (g(\hat{\theta}_{T_b,i}, \lambda) - \bar{g}(\lambda))^2 \tag{5.8}$$

where $\bar{g}(\lambda) = n^{-1} \sum_{i=1}^{n} g(\hat{\theta}_{T_b,i}, \lambda)$.

Under the conditions that $\{(\sqrt{T}(g(\hat{\theta}_T, \lambda) - g(\theta, \lambda)))^4\}$ are uniformly integrable, $n^{-1} \sum_{i=1}^{n} E|\sqrt{T_b}(g(\hat{\theta}_{T_b,i}, \lambda) - g(\theta, \lambda))|^2 \longrightarrow V(\theta, \lambda)$ as $n \rightarrow \infty$ and the proper mixing condition for $\{w(t)\}$, we have

$$\hat{V}(\lambda) \rightarrow V(\theta, \lambda) \text{ in } L^2 \text{ as } T \rightarrow \infty.$$

See Carlstein (1986) and Fukuchi (1999) for details of the variance estimation via subsampling.

Replacing T_b, n, $\sqrt{T_b}g(\hat{\theta}_{T_b,i}, \lambda)$, and $\sqrt{T_b}\bar{g}(\lambda)$ in the right-hand side of (5.8) with T_k, n_k, $\sqrt{T_k}M(\hat{\theta}_k^{(i)}, \lambda)$, and $\sqrt{T_k}\bar{M}_k(\lambda)$ provides $\hat{V}(\lambda) = \hat{H}_k(\lambda)$, which is the asymptotic variance estimator given in (5.6).

5.3   A Simulation Study of Finite Sample Test Properties

We observe the finite sample properties for the test statistics described in the previous section through a series of Monte Carlo simulations. Three examples are provided in this section: two for the simple and one for the partial causality changes. The results in Sects. 5.3 and 5.4 are generated using Ox version 7.10 (see Doornik 2013).

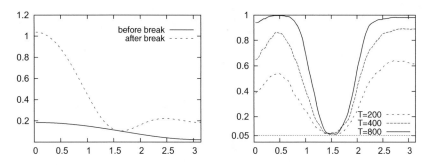

Fig. 5.1 True one-way effect measures (*left*) and rejection rates (*right*)

5.3.1   Change in Simple Causality Measure

Consider the following bivariate VARMA model with a structural break at $T_1 = 0.5T$.

$$\begin{cases} X(t) & = 0.3(1 + 1_{\{t > T_1\}}b(L))Y(t-1) + e_1(t), \\ Y(t) & = e_2(t) + 0.5e_2(t-1) \end{cases}$$

where $e_1(t)$ and $e_2(t)$ are independently distributed as a standard normal. $b(L) = 1 - 2\cos\lambda_0 L - L^2$ is the Gegenbauer polynomial, which is used in Breitung and Candelon (2006) to make an illustrative example for the non-causality test. The coefficient λ_0 in the Gegenbauer polynomial is set to 0.5π. The left panel of Fig. 5.1 shows the true one-way effect measures for the periods before ($t \leq T_1$) and after a break ($t > T_1$) using solid and dashed lines, respectively. In this illustration, the causality measure increases for all frequencies except $\lambda = \lambda_0$ for the period after the structural break. The variance estimate for the causality measure estimator is obtained through 100 Monte Carlo replications. The null hypothesis is that there is no causality change, and the nominal size of the test is set to 5%. The rejection rate of the null hypothesis is calculated using 1000 iterations. It is noted that the rejection rate corresponds to the power of the test statistic at frequency $\lambda \neq \lambda_0$ and the actual size at $\lambda = \lambda_0$. The right panel of Fig. 5.1 exhibits the powers and the actual sizes of the test statistic for sample sizes $T = 200$, 400, and 800. The rejection rate at $\lambda = \lambda_0$ is close to the nominal size 5%. The powers become higher as the departure from the null hypothesis becomes larger. However, the powers at near-zero frequencies are lower than those for other frequencies because the variances of the estimated measure at near-zero frequency are relatively large.

Next, we consider the testing procedure described in Sect. 5.2.1, which is the test for a change in measures using high-frequency data. We generate intra-daily data using following VAR(2) model

$$\begin{bmatrix} x(t) \\ y(t) \end{bmatrix} = A[1] \begin{bmatrix} x(t-1) \\ y(t-1) \end{bmatrix} + A[2] \begin{bmatrix} x(t-2) \\ y(t-2) \end{bmatrix} + \begin{bmatrix} \varepsilon_1(t) \\ \varepsilon_2(t) \end{bmatrix}, \quad \begin{bmatrix} \varepsilon_1(t) \\ \varepsilon_2(t) \end{bmatrix} \sim N(0, \Sigma).$$

(5.9)

The parameter values for $A[1]$, $A[2]$, and Σ are set to the estimated values using the intra-daily returns of Nikkei 225 index and its futures for a typical date, March 21, 2013. The model for $x(t)$ and $y(t)$ is as follows:

$$A[1] = \begin{bmatrix} -0.031 & 0.129 \\ 0.263 & -0.319 \end{bmatrix}, \quad A[2] = \begin{bmatrix} 0.064 & 0.086 \\ 0.193 & -0.133 \end{bmatrix}$$

(5.10)

and

$$\Sigma = 10^{-8} \begin{bmatrix} 2.32 & 1.40 \\ 1.40 & 3.81 \end{bmatrix}.$$

(5.11)

The size properties for the test statistic (5.7) are examined using the intra-daily data generated with model (5.9) with (5.10) and (5.11) for both subsample periods for which $n_1 = n_2 = 20$. The sample size of each group of intra-daily data is set to 100, 200, and 600. Using the generated intra-daily data of the ith day in subsample period k, we estimate $\theta_k^{(i)}$ for the model (5.9) and then compute the causality measures using the estimated parameters. Denote the estimated causality measure of the one-way effect from $\{y(t)\}$ to $\{x(t)\}$ for the ith day in subsample period k as $M(\hat{\theta}_k^{(i)}, \lambda)$, as in Sect. 5.2.1. The estimate of the causality measure for the subsample period k is given by $\bar{M}_k(\lambda)$ as in (5.5). The rejection rate for the null hypothesis that there is no change in the causality measures at frequency λ, which is $CM(\lambda) = 0$, is calculated using 1000 iterations. The actual size of the test statistic (5.7) is given as the rejection rate for the case in which we use the same parameter values for both subsample periods. The actual sizes of a test with a sample size of 100 for a given nominal of 5% exhibit small upward bias, which are at most 8% for all frequencies. The biases diminish as the sample size increases. For the test for a change in overall measures, we confirm that the actual size is 9% with a sample size of 100, which is almost the same amount as for the frequency-wise measure.

To see the power properties of the test statistic, we generate intra-daily data for subsample period 2 using the model (5.9) with (5.12) and (5.13) instead of (5.10) and (5.11), as follows

$$A[1] = \begin{bmatrix} -0.193 & 0.215 \\ -0.084 & 0.119 \end{bmatrix}, \quad A[2] = \begin{bmatrix} 0.1777 & -0.329 \\ -0.027 & -0.081 \end{bmatrix}$$

(5.12)

and

$$\Sigma = 10^{-8} \begin{bmatrix} 2.43 & 1.37 \\ 1.37 & 3.22 \end{bmatrix}.$$

(5.13)

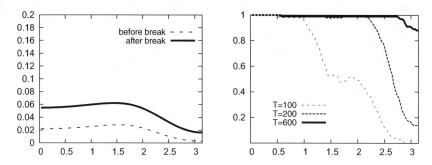

Fig. 5.2 True one-way effect measures (*left*) and rejection rates (*right*)

The parameter values in (5.12) and (5.13) are the estimated values for a typical date, April 24, 2013. The model (5.9) with (5.10) and (5.11) and the model (5.9) with (5.12) and (5.13) provide the true causality measures for the periods before and after a break, as shown in the left panel of Fig. 5.2. The power of the test statistic is given as the rejection rate for the case in which we use different parameter values for each subsample period. In the right panel of Fig. 5.2, rejection rates for each sample size are shown. The right panel indicates that the powers of the test statistic tend to unity with increasing sample size and become smaller with increasing related frequency. The latter is ascribed to the decreasing changes in causality measures with increasing frequency, as given in the left panel of Fig. 5.2. In addition to the frequency-domain measures, we examine the power properties of the test for overall measures, which are all unity for all sample sizes. This finding is consistent with the result of the frequency-domain measures in the sense that they both detect changes for all sample sizes.

5.3.2 Change in Partial Causality Measure

We examine the sample variation in the change estimates of partial measures using the sample average and MSE of Monte Carlo simulations, as in Sect. 4.2.2. For the partial measure estimation, a simulation study of the finite sample properties of tests is not conducted given high computational costs. The following trivariate VAR(2) model with a break point $T_1 = [0.5T]$ is considered:

$$\begin{bmatrix} x(t) \\ y(t) \\ z(t) \end{bmatrix} = A[1] \begin{bmatrix} x(t-1) \\ y(t-1) \\ z(t-1) \end{bmatrix} + A[2] \begin{bmatrix} x(t-2) \\ y(t-2) \\ z(t-2) \end{bmatrix} + \begin{bmatrix} \varepsilon_1(t) \\ \varepsilon_2(t) \\ \varepsilon_3(t) \end{bmatrix}$$

where

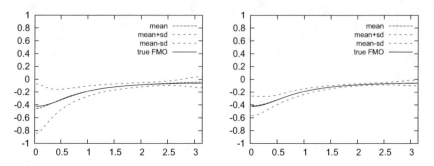

Fig. 5.3 One standard error confidence interval for changes in partial measures of Monte Carlo experiments: $T = 800$ (*left*) and $T = 3200$ (*right*)

$$A[1] = \begin{bmatrix} 0 & a & 0 \\ 0.6 & 0 & 0.5 \\ 0.6 & 0.6 & 0 \end{bmatrix}, A[2] = \begin{bmatrix} 0 & 0.3 & 0 \\ 0 & 0 & 0 \\ 0 & 0 & 0 \end{bmatrix}, a = 0.3 \cdot 1_{\{t > T_1\}}$$

and $(\varepsilon_1(t),\ \varepsilon_2(t),\ \varepsilon_3(t))' \sim N(0, I)$. In this experiment, sample size T is set to either 800 or 3200. Using this model, we estimate the change in partial causality measures from y to x. Figure 5.3 shows sample averages of the estimates of 100 Monte Carlo experiments and one standard error confidence intervals using different types of dashed lines, respectively. Also, the true measures are expressed with a solid line. As Fig. 5.3 shows, small biases exist in the low-frequency domain when $T = 800$, which disappeared in the case of $T = 3200$. Standard deviations decrease as the sample size increases and becomes relatively large in the low-frequency domain, as with the experiments in Sect. 4.2.2 (2).

5.4   Empirical Illustrations

In this section, illustrative empirical applications are provided. The first application is the predictability of aggregate stock returns using dividend yields. The analysis is conducted using the two subsample periods before and after the financial crisis of 2008. The second example is a change in the causal relationship between a stock index and its futures using high-frequency financial data. The daily causality measure is estimated using intra-daily data.

5.4.1   Stock Returns and Dividend Yields

We consider the predictability of aggregate stock returns using dividend yields. Numerous empirical studies investigated the predictability of stock returns.

In particular, dividend yields are candidates for predictors because of the present value relationship between dividends and stock prices. Campbell and Shiller (1988) is the well-known empirical study that investigates predictability using yearly US data and a VAR model. Recently, some empirical studies on the predictability of stock returns employ monthly data. For instance, Bollerslev et al. (2015) considered the dynamics of monthly returns, dividend yields, dividend growth rates, and variance risk premiums. Additionally, Aono and Iwaisako (2011) examined the predictability of stock returns using financial ratios, including the dividend yield, in the context of linear regressions.

In this study, we investigate the predictability of monthly returns of TOPIX. TOPIX is a stock index of the first sector of Tokyo Stock Exchange. As in Campbell and Shiller (1988), we estimate the time-series model with stock returns, dividend yields, and dividend growth rates. Predictability is evaluated by the partial one-way effect measures from dividend yields and the dividend growth rates, respectively, and simple measures from the vector that consists of dividend yields and the dividend growth rates. In the following, we call the latter a simple measure. We use monthly data on TOPIX and the dividend yields of stocks listed on the first sector of the Tokyo Stock Exchange from April 1999 to February 2017, as taken from the Nikkei NEEDS database.

Let $P(t)$ and $D(t)$ be a value of the stock index and the dividend paid on index, respectively. The return of the index, the dividend yield, and the dividend growth rate are computed as $r(t) = \log(P(t+1) + D(t)) - \log(P(t))$, $\delta(t) = \log(D(t)) - \log(P(t))$, and $g(t) = \log(D(t+1)) - \log(D(t))$, respectively. It is noted that the dividend yield $\delta(t)$ is detrended by subtracting the local trends for the periods before and after a break, respectively. Figure 5.4 indicates the series of returns, dividend yields, and dividend growth rates. The vertical dashed line represents September 2008, when Lehman Brothers went bankrupt. We call this month the break month and test the changes in the causal relationships across the event. The estimation procedure described in Sect. 4.1.1 is adopted to trivariate VARMA model for the process $\{r(t),\ \delta(t+1),\ g(t)\}$. At the Step A.2, second-order VAR models are selected for both before and after a break based on AIC. To see the changes in the causal relationships, we employ the procedures in Sect. 5.2.1.

Figure 5.5 depicts the partial frequency-wise measures of the one-way effect (FMO) from the dividend yields (left column), dividend growth rates (middle column), and simple FMO (right column). The top, middle, and bottom rows of Fig. 5.5 indicate the FMO, the change in FMO, and the test statistics for changes in FMO, respectively. The dashed lines in the bottom row of Fig. 5.5 depict the critical values of 5% significance for testing the null hypothesis that there is no change in causality.

The panels in the right column show that there are significant changes in the simple FMOs at a frequency larger than 0.4. In particular, FMOs are significantly strengthened in the high-frequency domain. The overall measures of the one-way effect (OMO) are 0.44 and 2.62 for the periods before and after a break, respectively, such that the change is −2.18. The change is statistically significant with the test statistic −3.17.

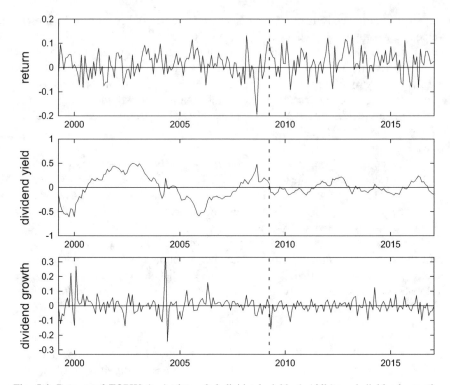

Fig. 5.4 Returns of TOPIX (*top*), detrended dividend yields (*middle*), and dividend growths (*bottom*)

In contrast, the change in partial measures of the dividend yield (left column) is not significant for all frequencies. The FMOs are strengthened in the high-frequency domain and weakened in the low-frequency domain. The OMOs are 0.53 and 0.31, respectively, such that the change is 0.22 with a test statistic of 0.14. This decrease is not significant.

The partial FMOs for dividend growth (middle column) are strengthened in all frequencies and significant in the very low-frequency domain. The OMO increases from 0.13 to 1.95 across the break but not significantly, with a test statistic of −1.62.

The significant increase in the simple measures implies an improvement in predictability. However, the source of the improvement is ambiguous because the changes in partial measures are not significant in most frequency domains. It is noted that the one-way effects from the dividend yield and growth rate are not only possible sources of improvement but also their reciprocal component.

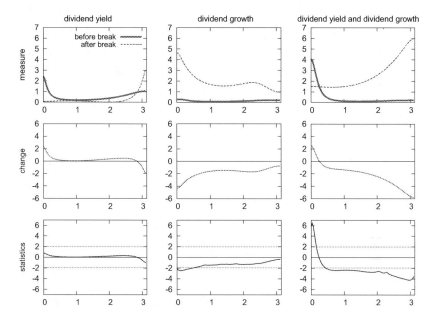

Fig. 5.5 Causality measures (*top row*), change in measures (*middle*), and test statistics (*bottom*)

5.4.2 Intra-Daily Financial Time Series

We consider a change in the causal relationship between the stock index and its futures using high-frequency financial data. Financial theory implies that there is no predictability between a stock index and its futures because they are based on almost the same information in efficient financial markets. However, many empirical studies showed evidence for predictability, especially in intra-daily variations. In the context of Granger causality, Abhyankar (1999) reports on the predictability from futures to spot returns using the FTSE and its futures. The predictabilities are found in both directions using a nonlinear model, whereas the predictability from the futures was found using the linear model. Fleming et al. (1996), in their empirical study using S&P 500 and its futures, pointed out that transaction costs are relatively small for the futures market and found predictability from the futures to the stock index. Yang et al. (2012) found that the stock index leads futures for the Chinese stock market when the futures market is in its infancy.

Some market microstructure theories suggest predictability from futures returns to spot returns. For example, Kyle (1985) shows that a trader who has private information would like to trade in a market with large liquidity to ensure a weak price impact. In this context, the futures market has predictability for the stock index because it has large liquidity. Diamond and Verrecchia (1987) showed that a short-selling regulation slows down the convergence speed of spot prices to the fundamental value. This result also implies predictability from futures returns to spot returns.

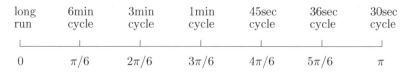

long run	6min cycle	3min cycle	1min cycle	45sec cycle	36sec cycle	30sec cycle
0	$\pi/6$	$2\pi/6$	$3\pi/6$	$4\pi/6$	$5\pi/6$	π

Fig. 5.6 Correspondence between frequencies and cycles

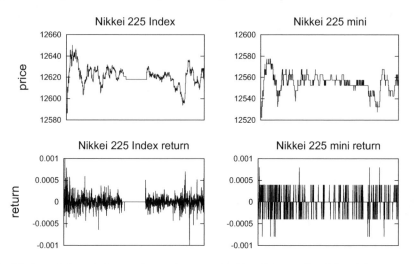

Fig. 5.7 Intra-daily variation of prices and returns of Nikkei 225 and Nikkei 225 mini

Now, we consider a causal relationship between the Nikkei 225 and the Nikkei 225 mini. The Nikkei 225 mini is a futures contract on the value of the Nikkei 225. We use 15 seconds of log returns for these series. The correspondence between frequencies and cycles when we use the 15-second interval data is shown in Fig. 5.6. To compute the log returns, we employ futures prices with the nearest maturity as representative prices and use a mid-price, which is the average of the bid and ask quotes. We separately analyze the intra-daily series for the morning session (9:00 to 11:30) and the afternoon session (12:30 to 15:00) because the Tokyo Stock Exchange spot market has a lunch break from 11:30 to 12:30. The intra-daily series of a typical date, March 21, 2013, is shown in Fig. 5.7. The Nikkei 225 mini varies discretely because its tick size is 5 Japanese yen, whereas the Nikkei 225 behaves as a continuous variable because it is an average of individual stock prices.

As explained in Sect. 5.2.2, we estimate the VAR model on every business day using intra-daily data. The VAR model lags are determined by the AIC for each day. Using the estimated parameters, the causality measures are computed for each day. In addition to causality measure analysis, we investigate the causality changes across the announcement of a monetary policy. The Bank of Japan announced on April 4, 2013, the introduction of Quantitative–Qualitative Easing (QQE). We divide the entire sample period (February 25 to May 16, 2013) into three periods: the pre-event

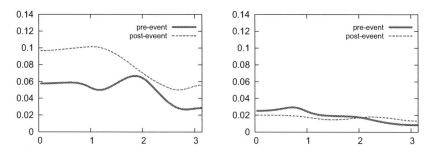

Fig. 5.8 One-way effect measures for pre- and post-event periods in the morning session: futures to index (*left*) and index to futures (*right*)

period (February 25 to March 25), the event period (March 26 to April 15), and the post-event period (April 16 to May 16). We consider the pre- and post-event periods as the periods before and after a break, respectively. We conduct a test of whether there are changes in the causality measures for the pre-event and post-event periods. The causality measures for the pre-event and post-event periods are estimated as the sample average of the estimates of each day, as in (5.5).

The average FMOs from the futures to the stock index are depicted in the left panel and the reverse direction is in the right panel of Fig. 5.8. We can see that the FMO from the futures to the stock index peaks at a frequency of approximately 1.8 for the pre-event period, and that the peak shifts to near the frequency 1.2 for the post-event period. This result implies that the component with a 50-sec cycle is mainly reflected in the Nikkei 225 mini in advance of the Nikkei 225 for the pre-event period, and the 1.3-min cycle is a main component for the post-event period. We also find that the FMOs increase for all frequencies after the announcement of QQE. In contrast, the FMOs from the stock index to the futures are small for all frequencies, as shown in the right panel of Fig. 5.8. Therefore, we focus on the causality measures from the futures to the stock index and their changes in the following.

Figure 5.9 shows the transitions of the day-by-day FMO estimates, more concretely at frequencies $\pi/3$ and $2\pi/3$, respectively. In the figure, the dashed lines

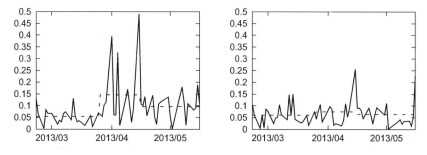

Fig. 5.9 Transitions of FMO estimates in the morning session at $\pi/3$ (*left*) and at $2\pi/3$ (*right*)

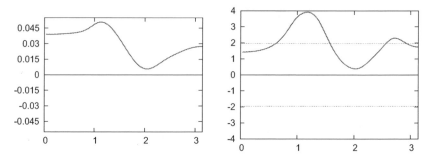

Fig. 5.10 Change in causality (*left*) and test statistics (*right*) in the morning session

represent the averages of the FMOs for the pre-event, event, and post-event periods. The change in the causality measures, defined as the difference between the average causality measure for the post-event period and that for the pre-event period, is shown in the left panel of Fig. 5.10. The test statistic for the non-causality change and the critical values for the 5% significance level are depicted by solid and dashed lines in the right panel. As shown in Fig. 5.9, the FMOs are relatively small for the pre-event period and become volatile in the event period at frequency $\pi/3$. For the post-event period, the FMOs tend to be large compared with those for the pre-event period, and the increase is statistically significant, as shown in Fig. 5.10. In contrast, the small increase at frequency $2\pi/3$ is not significant. The changes are significant at frequencies from approximately 0.6–1.5, which correspond to one-minute to 2.6-minute cycles. The OMOs from the futures to the stock index are 0.051 and 0.081 for the pre- and post-event periods, and the test statistic for the causality change is 0.5, such that it is not significant despite the changes being detected in view of the FMO.

For the afternoon session, the average FMOs are shown in Fig. 5.11. Those FMOs are increased except for very high frequency across the event. A peak in the FMO shifts from a frequency of approximately 1.0 to 0.8, which corresponds to 1.5- and

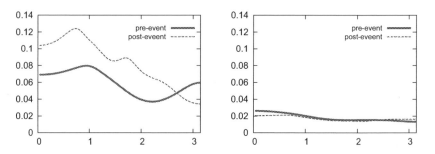

Fig. 5.11 One-way effect measures for pre- and post-event periods in the afternoon session: futures to index (*left*) and index to futures (*right*)

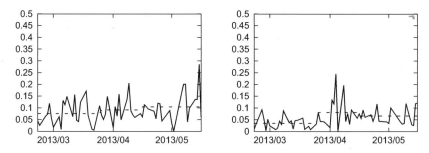

Fig. 5.12 Transitions of FMO estimates in the afternoon session at $\pi/3$ (*left*) and at $2\pi/3$ (*right*)

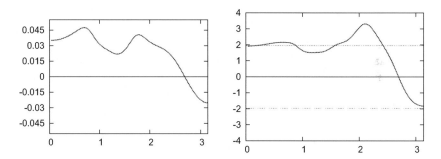

Fig. 5.13 Change in causality (*left*) and test statistics (*right*) in the afternoon session

2-minute cycles. The FMOs from the stock index to the futures are as small as the morning session. Figure 5.12 shows the transition of the day-by-day FMO estimates at frequencies $\pi/3$ and $2\pi/3$ and the averages of each period. The increment of the average of the causality measures and the test statistics for causality changes are shown in Fig. 5.13. Although Fig. 5.12 indicates that the averages of the FMO increase across the event both at frequency $\pi/3$ and $2\pi/3$, the increase is significant at $2\pi/3$ but not at $\pi/3$, as in Fig. 5.13. The OMOs from the futures to the stock index are 0.059 and 0.083 for the pre- and post-event periods, respectively, and the test statistic is 0.41, such that a causality change is not significant as in the morning session.

To summarize, we find that there are some predictabilities from the Nikkei 225 mini to the Nikkei 225 in both morning and afternoon sessions. These predictabilities strengthened in most frequencies after the announcement of the Quantitative–Qualitative Easing by Bank of Japan. The changes in FMO are statistically significant in some frequency domains, and the changes in OMO are not significant in both morning and afternoon sessions.

References

Abhyankar, A. (1999). Linear and nonlinear Granger causality: Evidence from the U.K. stock index futures market. *Journal of Futures Markets*, *18*(5), 519–540.

Andrews, D. W. K. (1993). Tests for parameter instability and structural change with unknown change point. *Econometrica*, *61*(4), 821–856.

Aono, K., & Iwaisako, T. (2011). Forecasting Japanese stock returns with financial ratios and other variables. *Asia-Pacific Financial Markets*, *18*(4), 373–384.

Bollerslev, T., Xu, L., & Zhou, H. (2015). Stock return and cash flow predictability: The role of volatility risk. *Journal of Econometrics*, *187*(2), 458–471.

Breitung, J., & Candelon, B. (2006). Testing for short- and long-run causality: A frequency-domain approach. *Journal of Econometrics*, *132*(2), 363–378.

Campbell, J. Y., & Shiller, R. J. (1988). Stock prices, earnings, and expected dividends. *Journal of Finance*, *43*(3), 661–676.

Carlstein, E. (1986). The use of subseries values for estimating the variance of a general statistic from a stationary sequence. *The Annals of Statistics*, *14*(3), 1171–1179.

Diamond, D. W., & Verrecchia, R. E. (1987). Constraints on short-selling and asset price adjustment to private information. *Journal of Financial Economics*, *18*(2), 277–311.

Doornik, J. A. (2013). *Object-oriented matrix programming using Ox* (3rd ed.). London: Timberlake Consultants Press and Oxford. www.doornik.com.

Fleming, J., Ostdiek, B., & Whaley, R. E. (1996). Trading costs and the relative rates of price discovery in stock, futures, and option markets. *Journal of Futures Markets*, *16*(4), 353–387.

Fukuchi, J. (1999). Subsampling and model selection in time series analysis. *Biometrika*, *86*(3), 591–604.

Hansen, B. E. (1996). Inference when a nuisance parameter is not identified under the null hypothesis. *Econometrica*, *64*(2), 413–430.

Hosoya, Y. (1997). A limit theory for long-range dependence and statistical inference on related models. *The Annals of Statistics*, *25*, 105–137.

Kyle, A. S. (1985). Continuous auctions and insider trading. *Econometrica*, *53*(6), 1315–1335.

Yang, J., Yang, Z., & Zhou, Y. (2012). Intraday price discovery and volatility transmission in stock index and stock index futures markets: Evidence from China. *Journal of Finance*, *32*(2), 99–121.

Appendix
Technical Supplements

Abstract The appendix consists of three sections dealing with Hilbert space, a root contraction method of AR coefficients and the frequency-domain Whittle likelihood function, respectively. Section A.1 gives a short description of Hilbert space. Section A.2 uses the Jordan decomposition to introduce an eigenvalue contraction algorithm that guarantees the multivariate ARMA coefficient estimates to satisfy a set of predetermined root conditions. The Whittle likelihood is the approximate Gaussian likelihood proposed by Whittle (1952). Section A.3 provides the Whittle likelihood function for the cointegrated vector ARMA process, which includes the stationary process as a special case.

Keywords Jordan form, root conditions, root contraction method, Whittle likelihood function.

A.1 Hilbert Space

This section is for a brief introduction to real Hilbert space. A real Hilbert space H is a linear (vector) space so that if $x_1, x_2 \in H$, then $\alpha_1 x_1 + \alpha_2 x_2 \in H$ if α_1 and α_2 are real numbers; namely, it is a set closed under linear combination. Formally, a real linear space H is said to be a Hilbert space if it has the following two properties:

(1) An inner product (x_1, x_2) is defined for any elements $x_1, x_2 \in H$ where the inner product is a bilinear operation satisfying (i) $(x_1, x_2) = (x_2, x_1)$, (ii) $(\alpha_1 x_1 + \alpha_2 x_2, x_3) = \alpha_1 (x_1, x_3) + \alpha_2 (x_2, x_3)$. The inner product induces the norm $||x|| \equiv \sqrt{(x, x)}$ on any $x \in H$.
(2) The space H is complete in respect of the norm. Namely, if $\{x_i, i = 1, 2, \dots\}$ is a sequent of elements of H satisfying $\lim_{i,j \to \infty} ||x_i - x_j|| = 0$, then there is an element of $x \in H$ to which the sequence $\{x_i\}$ converges in the sense that $\lim_{i \to \infty} ||x_i - x|| = 0$. The sequence satisfying the condition $\lim_{i,j \to \infty} ||x_i - x_j|| = 0$ is said a Cauchy sequence.

Simple examples of real Hilbert space are the set of all real numbers with the inner product given by the product $(x_1, x_2) = x_1 x_2$ and the set of the real n-dimensional vectors whose inner product is defined by $(x_1, x_2) \equiv \sum_{k=1}^{n} x_{1,k} x_{2,k}$ where $x_{1,k}$

© The Author(s) 2017
123
Y. Hosoya et al., *Characterizing Interdependencies of Multiple Time Series*,
JSS Research Series in Statistics, DOI 10.1007/978-981-10-6436-4

and $x_{2,k}$ are the k-th components of the real vectors x_1, x_2. Another example of finite dimensional vector space is the space generated by linear combinations of (spanned by) p real-valued random variables $\{x_1, x_2, \ldots, x_p\}$ with mean zero and finite variance where the inner product is given by the expectation of the product $(x_1, x_2) \equiv E(x_1 x_2)$. To express explicitly the generating variables, we write $H_1 \equiv H\{x_1, x_2, \ldots, x_p\}$. Let $H_2 \equiv H\{x_2, x_3, \ldots, x_p\}$ be another space spanned by $x_2, x_3, \ldots, x_p$. Evidently $H_2 \subset H_1$ and H_2 is said a subspace of H. But in the case of infinite dimensional space where $p = \infty$, we need to take into account the limits of sequences. If $p = \infty$, the Hilbert space H_1 generated by the infinite sequence $\{x_1, x_2, \ldots\}$ is defined to be the (topological) closure of the space spanned by elements of $\{x_1, x_2, \ldots\}$ in the Hilbert space of all random variables of finite variance, namely all the limits of Cauchy sequences of generated by linear combinations of $x_1, x_2, \ldots$ are included in the space H_1. In the same way, the subspace $H_2 \equiv H\{x_2, x_3, \ldots\}$ is defined to be the completion of the linear space spanned by the sequence $x_2, x_3, \ldots$ The projection of x_1 on H_2 is the element $z \in H_2$ such that $Var(x_1 - z) = \min_{y \in H_2} Var(x_1 - y)$. There is such a unique element since H_2 is a closed subspace. The difference $x_1 - z$ is termed the residual of the projection. For more detailed explanation of the concept of Hilbert space and its application, see Rozanov (1967), Hannan (1970), and Brillinger (1975). They provide detailed explanation of the derivation of the moving average representation of a vector stationary process from a canonical factorization of spectral density matrix in the frame work of Hilbert space representation of the second-order stationary process. The derivation is frequently used in Chaps. 2 and 3 of this book, and Sect. 2.4.2 of this book alludes to the correspondence.

A.2   Root Contraction Method

Let a p-vector MA process $\{\xi(t)\}$ be generated by $\xi(t) = \sum_{l=0}^{b} B[l]\varepsilon(t - l)$, where $B[0] = I_p$. The $\varepsilon(t)$s are assumed white noise vectors with mean 0 and covariance matrix $\Sigma^\dagger$; they determine the second-order properties of the process $\{W_T(t)\}$ generated, for $t = 1, \cdots, T$, by

$$\Delta(L)W_T(t) = \Pi W_T(t-1) + \sum_{k=1}^{a-1} \Gamma[k]\Delta(L)W_T(t-k) + \Gamma g_T(t) + \xi(t), \quad (\text{A.1})$$

where $\Pi = \alpha\beta'$ for $p \times r$ matrices α, β, $\Delta(L) = (1 - L)I_p$, $g_T(t)$ is a deterministic s-vector involving no unknown parameters, and Γ is a $p \times s$ coefficient parameter. Assume that the $W_T(t)$ for $-a + 1 \leq t \leq 0$ are uniformly bounded in probability with respect to T. Set

$$A(z) \equiv \Delta(z) - \Pi z - \sum_{k=1}^{a-1} \Gamma[k] z^k \Delta(z).$$

Let $(\beta, \beta_\perp)$ be a $p \times p$ matrix such that $\beta \perp \beta_\perp$ and $rank(\beta, \beta_\perp) = p$. We focus on the model of cointegration rank $rank\beta = r$; then, we have the relationship:

$$A(z)[\beta, \beta_\perp] = [A(z)\beta, \beta_\perp - \sum_{k=1}^{a-1} \Gamma[k] z^k \beta_\perp] \left[\begin{array}{c|c} I_r & 0 \\ \hline 0 & (1-z)I_{p-r} \end{array} \right]$$

$$\equiv C(z) \left[\begin{array}{c|c} I_r & 0 \\ \hline 0 & (1-z)I_{p-r} \end{array} \right] \equiv \sum_{k=0}^{a} C_k z^k \left[\begin{array}{c|c} I_r & 0 \\ \hline 0 & (1-z)I_{p-r} \end{array} \right]. \quad (A.2)$$

Therefore, all roots of $\det C(z) = 0$ are outside the unit circle if and only if $(p-r)$ roots of $\det A(z) = 0$ are ones and the rest are outside the unit circle.

The root conditions: Under the cointegration rank r hypothesis, the characteristic polynomial satisfies the condition that $\det C(z) = 0$ only if $|z| > 1$. Moreover, all roots of $\det B(z) = 0$ are assumed to be on or outside the unit circle and do not share any common zero with $\det A(z)$.

To be explicit, the matrix $C(z)$ in (A.2) is represented as, if $a = 1$,

$$C(z) = [\beta, \beta_\perp] - [(I_p + \Pi)\beta, 0]z,$$

whereas, if $a \geq 2$,

$$C(z) = [\beta, \beta_\perp] - [(I_p + \Pi + \Gamma[1]\beta, \Gamma[1]\beta_\perp]z$$

$$- 1\{a \geq 3\} \sum_{k=2}^{a-1} [(\Gamma(k) - \Gamma[k-1])\beta, \Gamma[k]\beta_\perp]z^k + [\Gamma[a-1]\beta, 0]z^a,$$

where the bracketed matrix pairs consist of $p \times r$ and $p \times (p-r)$ matrices. Suppose at first that $a \geq 2$; we set $D_a \equiv \Gamma[a-1]\beta$ and define by J_r the $r \times p$ matrix whose (i, j) component is 1 if $i = j$ and 0 otherwise. The companion matrix for the generating mechanism $C(z)$ is defined by the following $q \times q$ square matrix (where $q \equiv (a-1)p + r$)

$$D \equiv \left[\begin{array}{ccccc} -C_0^{-1}C_1 & -C_0^{-1}C_2 & \cdots & -C_0^{-1}C_{a-1} & -C_0^{-1}D_a \\ I_p & 0 & \cdots & \cdots & 0 \\ 0 & I_p & \vdots & \ddots & \vdots \\ 0 & 0 & \cdots & J_r & 0 \end{array} \right].$$

Let v_i and ω_i, $i = 1, \cdots, q$, be the eigenvectors and eigenvalues of D; namely,

$$D = [v_1, \cdots, v_q] diag(\omega_1, \cdots, \omega_q)[v_1, \cdots . v_q]^{-1}.$$

Furthermore, let $v_i^{(j)}$ be the partitions of the eigenvector v_i such that the $v_i^{(j)}$ are p-vectors for $1 \le j \le a - 1$ and $v_i^{(a)}$ is a r-vector. Hence, $v_i = (v_i^{(1)'}, \cdots, v_i^{(a)'})'$. Then, for each i, we have, as long as $|\omega_i| \ne 0$,

$$\left(I + C_0^{-1} \sum_{k=1}^{a-1} C_j \omega_i^{-k} + C_0^{-1} D_a J_r \omega_i^{-a}\right) v_i^{(1)} = 0,$$

$$v_i^{(j)} = \omega_i^{-j+1} v_i^{(1)}, \quad j = 2, \cdots, a - 1; \quad v_i^{(a)} = \omega_i^{-a+1} J_r v_i^{(1)}, \quad j = a. \quad \text{(A.3)}$$

In the sequel, we assume without much loss of generality that all nonzero eigenvalues $v_1, \ldots, v_q$ are distinct. The roots of $\det C(z) = 0$ are nothing but the inverse of the eigenvalues of the matrix D, such that the roots condition of $C(z)$ is equivalent to all ω_is being inside the unit circle [see Miller (1968), p.37.]. We sometimes encounter such cases as some ω_js for D in conventional estimation procedures of (A.1) falling on or outside the unit circle, violating the assumption of $rank(\Pi) = r$. For example, the C_js induced from the unrestricted ML estimate might produce such an eigenvalue set. To test cointegration rank r, $r = 0, 1, \cdots, p - 1$ against rank p on the basis of LR statistics, we must prepare all of the likelihoods, where each likelihood should be estimated with a specified rank of Π. A way to modify the model parameters to satisfy the condition of stationarity is as follows. If the ω_i, $i = 1 \cdots, s$, are eigenvalues such that $|\omega_i| > 1 - \varepsilon_1$ for a suitably chosen small number ε_1 and $|\omega_1| > |\omega_2| > \cdots > |\omega_s|$, contract them to $\omega_1^\dagger, \cdots, \omega_s^\dagger$ such that $|\omega_1^\dagger| > |\omega_2^\dagger| > \cdots > |\omega_s^\dagger|$, $|\omega_{s+1}| \le |\omega_i^\dagger| \le 1 - \varepsilon_1$ and $\arg(\omega_i^\dagger) = \arg(\omega_i)$, $i = 1, \cdots, s$, where ω_{s+1} is the eigenvalue such that $|\omega_{s+1}| = \max_{s+1 \le j \le q} |\omega_j|$. A way to do this is to set

$$\omega_i^\dagger \equiv \frac{\omega_i}{|\omega_i|} \left\{ |\omega_{s+1}| + \frac{(s - i + 1)(1 - \varepsilon_1 - |\omega_{s+1}|)}{s} \right\}. \quad \text{(A.4)}$$

In this way, we contract the roots on or outside the $(1 - \varepsilon_1)$ circle for the correction to be minimal and contract them without distorting the order with respect to the absolute value. Using those modified $\omega_i^\dagger$s, define $D^\dagger$ by

$$D^\dagger \equiv [v_1^\dagger, \cdots, v_s^\dagger, v_{s+1}, \cdots, v_q] diag(\omega_1^\dagger, \cdots, \omega_s^\dagger, \omega_{s+1}, \cdots, \omega_q)$$

$$\cdot [v_1^\dagger, \cdots, v_s^\dagger, v_{s+1}, \cdots, v_q]^{-1},$$

where the new eigenvectors $v_i^\dagger$, $i = 1, \cdots, s$, are given, in parallel to (A.3), by

$$v_i^{(1)\dagger} = v_i^{(1)}; \quad v_i^{(j)\dagger} = (\omega_i^\dagger)^{-j+1} v_i^{(1)}, \quad j = 2, \cdots, a - 1; \quad v_i^{(a)\dagger} = (\omega_i^\dagger)^{-a+1} J_r v_i^{(1)}.$$

By this modification, $D^\dagger$ is given as

$$
D^\dagger = \begin{bmatrix}
-C_0^{-1}C_1^\dagger & -C_0^{-1}C_2^\dagger & \cdots & -C_0^{-1}C_{a-1}^\dagger & -C_0^{-1}D_a^\dagger \\
I_p & 0 & \cdots & 0 & 0 \\
0 & I_p & \ddots & \vdots & \vdots \\
0 & 0 & \ddots & 0 & 0 \\
0 & 0 & \cdots & J_r & 0
\end{bmatrix},
$$

whose eigenvectors and eigenvalues are given by $v_1^\dagger, \cdots, v_s^\dagger, v_{s+1}, \cdots, v_q$ and $\omega_1^\dagger, \cdots, \omega_s^\dagger, \omega_{s+1}, \cdots, \omega_q$, respectively. Based on the elements of $D^\dagger$, we can produce the desirable AR coefficients whose characteristic roots are outside the unit circle.

In the case of $a = 1$, the companion matrix D is defined by $D = C_0^{-1}C_1$. If we denote the $r \times r$ upper left block of $C_0^{-1}C_1$ by $(C_0^{-1}C_1)^{(1,1)}$, the nonzero eigenvalues of D are given by the roots of $\det\{\omega I_r - (C_0^{-1}C_1)^{(1,1)}\} = 0$. Suppose that $\omega_1, \cdots, \omega_r$ are those nonzero eigenvalues (hence $\omega_{r+1} = \cdots = \omega_p = 0$) and are distinct. If the absolute values of the eigenvalues $\omega_i, i = 1, \cdots, s$, are greater than $(1 - \varepsilon_1)$, then define the contracted $\omega_i^\dagger$s as in (A.4), such that the modified $D^\dagger$ is given by

$$
D^\dagger = (v_1, \cdots, v_p)\, diag(\omega_1^\dagger, \cdots, \omega_s^\dagger, \omega_{s+1}, \cdots, \omega_p)(v_1, \cdots, v_p)^{-1}.
$$

The change of the coefficients C_i to $C_i^\dagger$ produces the new parameters $\alpha^\dagger$, $\Gamma(k)^\dagger$, $k = 1, \cdots, a - 1$. Namely, setting $C_i^\dagger = [C_i^{(1)\dagger}, C_i^{(2)\dagger}]$ where $C_i^{(1)\dagger}$ and $C_i^{(2)\dagger}$ are $p \times r$ and $p \times (p - r)$ matrices, for given β and $\beta_\perp$, we have

$$
\Gamma[a - 1]^\dagger = [C_a^{(1)\dagger}, C_{a-1}^{(2)\dagger}](\beta, -\beta_\perp)^{-1};
$$
$$
\Gamma[k - 1]^\dagger = [C_k^{(1)\dagger} + \Gamma[k]^\dagger\beta, C_{k-1}^{(2)\dagger}][\beta, -\beta_\perp]^{-1} \quad \text{for} \quad 2 \le k \le a - 1;
$$
$$
\alpha^\dagger = -(C_1^{(1)\dagger} - \beta - \Gamma[1]^\dagger\beta)(\beta'\beta)^{-1}.
$$

Consequently, we have a new set of AR coefficients whose Jordan-form eigenvalues are inside the unit circle. The validity of the root condition for the MA coefficient estimate $B(z)$ is also examined, and if some eigenvalues of the Jordan form are outside the unit circle, we modify them in a parallel way such that all of them are inside the unit circle. However, for MA coefficients, there is a better modification approach of canonical factorization of the MA spectral density matrix.

A.3   The Whittle Likelihood Function

Whittle (1952) proposed an approach to approximate the likelihood for observations generated from a scalar-valued stationary Gaussian process with parametric spectral density containing a finite set of unknown parameters. This section presents a type of Whittle likelihood function for the cointegrated vector ARMA process, which includes the stationary process as a special case.

In general, suppose that a p-vector process $\{W_T(t)\}$, $t = -a + 1, ..., T$, is generated as in (A.1) by the following error-correction form:

$$\Delta(L)W_T(t) = \Pi W_T(t-1) + \sum_{k=1}^{a-1} \Gamma[k]\Delta(L)W_T(t-k) + \Gamma g_T(t) + \xi(t), \quad \text{(A.5)}$$

where $\xi(t) = \sum_{l=0}^{b} B[l]\varepsilon(t-l)$ and under the hypothesis $H_r : rank(\Pi) = r$, Π is expressed as $\Pi = 0$ if $r = 0$, $\Pi = \alpha\beta'$ for full-rank $p \times r$ matrices α and β if $1 \leq r \leq p - 1$ and $\Pi = \alpha$ if $r = p$. The matrix β is assumed to be known. The deterministic s-vector $g_T(t)$ can contain a constant or a constant plus trend terms. Denote by $f_{\xi\xi}$ the spectral density of the MA process $\xi(t)$; namely,

$$f_{\xi\xi}(\lambda) = \frac{1}{2\pi} B(\lambda) \Sigma^{\dagger} B(\lambda)^{*},$$

where $B(\lambda) \equiv \sum_{l=0}^{b} B[l]e^{il\lambda}$. The Whittle likelihood uses the full sample data, exploiting the assumption that $\Delta(L)Z_T(t)$ and $\hat{\beta}'Z_T(t-1)$ are stationary under H_r. Set

$$W_1(\lambda) = (2\pi T)^{-1/2} \left\{ I_p - \sum_{j=1}^{a-1} \Gamma(j)e^{-ij\lambda} \right\} \sum_{t=2}^{T} e^{it\lambda} \Delta(L)Z_T(t), \quad \text{(A.6)}$$

$$W_2(\lambda) = (2\pi T)^{-1/2} \sum_{t=2}^{T} e^{it\lambda} ((\beta' Z_T(t-1))', g_T'(t))', \quad \text{(A.7)}$$

$$\psi = (vec(\alpha)', vec(\Gamma)')', \quad \text{(A.8)}$$

where $W_1(\lambda)$, $W_2(\lambda)$, and ψ are p, $(r+s)$, and $p(r+s)$-vector functions, respectively. The Whittle likelihood is defined as the function $L(Q(\psi, \Sigma^{\dagger(a,b)})) \equiv -T/2 \times Q(\psi, \Sigma^{\dagger(a,b)})$, where $Q(\psi, \Sigma^{\dagger(a,b)})$ is a real-valued function defined by

$$Q(\psi, \Sigma^{\dagger(a,b)}) = \log \det \Sigma^{\dagger(a,b)} + \frac{1}{2\pi} \int_{-\pi}^{\pi} \{W_1(\lambda) - (W_2(\lambda)' \otimes I_p)\psi\}^{*}$$
$$\cdot f_{\xi\xi}(\lambda)^{-1} \{W_1(\lambda) - (W_2(\lambda)' \otimes I_p)\psi\} d\lambda$$

where $\psi = (vec(\alpha)', vec(\Gamma(1))', \cdots, vec(\Gamma(a-1))', vec(\Gamma)')'$ is a $p \times (r + p \times (a-1) + s)$-vector. The Whittle likelihood is the function $L(Q(\psi, \Sigma^{\dagger(a,b)})) \equiv -T/2 \times Q(\psi, \Sigma^{\dagger(a,b)})$. The minimizing value of $\Sigma^{\dagger(a,b)}$ of Q is given by

$$\hat{\Sigma}^{\dagger}{}^{(a,b)} = \int_{-\pi}^{\pi} Re[\{B(\lambda)^{-1}W_1(\lambda) - (W_2(\lambda)' \otimes B(\lambda)^{-1})\psi\}$$
$$\cdot \{B(\lambda)^{-1}W_1(\lambda) - (W_2(\lambda)' \otimes B(\lambda)^{-1})\psi\}^*]d\lambda$$

for which holds the relation

$$Q(\psi, \hat{\Sigma}^{\dagger}{}^{(a,b)}) = \log \det \hat{\Sigma}^{\dagger}{}^{(a,b)} + p.$$

After obtaining the maximum Whittle likelihood estimates for each combination of the ARMA orders (a, b), we could select the lag structure of the ARMA model using, for example, the BIC, which is given by

$$BIC(a, b) = \log \det \hat{\Sigma}^{\dagger} + p \cdot \log \frac{1}{2\pi} + p + \frac{[pr + p^2\{(a - 1) + b\} + s]\log T}{T},$$

for $a \leq \bar{a}$ and $b \leq \bar{b}$, where $\bar{a}$ and $\bar{b}$ are chosen beforehand.

Remark A.1 Sect. 4.1 explains a three-step estimation procedure in which the third step is the maximum Whittle estimation. For computations in Chaps. 4 and 5, setting the coefficient estimate obtained in Step 2 in the three-step method as the initial value, we solve the minimizing problem of $\log \det \hat{\Sigma}^{\dagger}$ for the parameter vector ψ using a quasi-Newton method. Because, in general, the ARMA parameter ψ is not identifiable without some additional restrictions, it is more proper to set $\psi = \psi(\theta)$ for some identifiable parameter θ. For the likelihood to be maximized in the time-domain representation, it is necessary to evaluate residual sequences in each iteration step to estimate the covariance matrix $\Sigma^{\dagger}$. In contrast, the Whittle approach reduces the computational amount by omitting the exploitation of residuals. The root modification method in Step 2 places the initial estimate in the admissible set. To keep the coefficient estimate inside the admissible set during the optimization iteration, we may add two penalty terms to the objective function $\log \det \hat{\Sigma}^{\dagger}$. For a suitably small positive ε_2 and a suitably fixed positive value d, define a new objective function by

$$\log \det \hat{\Sigma}^{\dagger} + 1\{\max |\omega_{i_1}| > 1 - \varepsilon_2\}de^{-1/u_1} + 1\{\max |\omega_{i_2}| > 1 - \varepsilon_2\}de^{-1/u_2},$$

where ω_{i_1} and ω_{i_2} are eigenvalues of the companion matrix for the AR and MA parts, respectively, $i_1 = 1, \cdots, (a - 1)p + r$ and $i_2 = 1, \cdots, bp$, and $u_j = \max |\omega_{i_j}| - (1 - \varepsilon_2)$. Namely, u_j expresses the distance between the maximal eigenvalue and unity. Therefore, the further $u_j > 0$ is away from zero, the heavier the penalty e^{-1/u_j} becomes. If u_j is negative, all of the eigenvalues are inside the unit circle, and these penalty terms do not work. The constant d should be chosen such that $\log \det \Sigma^{\dagger}$ and the penalties are comparable magnitudes.

References

Brillinger, D. R. (1975). *Time Series Data Analysis and Theory*. New York: Holt, Rinehart and Winston Inc.

Hannan, E. J. (1970). *Multiple Time Series*. New York: John Wiley & Sons Inc.

Miller, K. S. (1968). *Linear Difference Equations*. New York: W.A. Benjamin Inc.

Rozanov, Y. A. (1967). *Stationary Random Processes*. San Francisco: Holden Day.

Whittle, P. (1952). Some results in time series analysis. *Skandinavisk Aktuarietidskrift, 1–2*, 48–60.

Index

© The Author(s) 2017   131
Y. Hosoya et al., *Characterizing Interdependencies of Multiple Time Series*,
JSS Research Series in Statistics, DOI 10.1007/978-981-10-6436-4

Printed in the United States
By Bookmasters